T0234469

Unsaturated Soil Mechanics with Probability and Statistics

Unsaturated Soil Mechanics with Probability and Statistics

Ryosuke Kitamura and Kazunari Sako

CRC Press
Taylor & Francis Group
Boca Raton London New York

CRC Press is an imprint of the
Taylor & Francis Group, an **informa** business

CRC Press
Taylor & Francis Group
6000 Broken Sound Parkway NW, Suite 300
Boca Raton, FL 33487-2742

First issued in paperback 2021

ISBN-13: 978-1-138-55368-2 (hbk)
ISBN-13: 978-1-03-217748-9 (pbk)
DOI: 10.1201/9780429320132

Library of Congress Control Number: 2019943569

Visit the Taylor & Francis Web site at
http://www.taylorandfrancis.com

and the CRC Press Web site at
http://www.crcpress.com

Contents

Preface

Unsaturated Soil Mechanics with Probability and Statistics is a theoretical textbook on unsaturated coarse-grained soil aided by probability theory and statistics, in which microscopic consideration is given to analyzing the mechanical behavior of unsaturated coarse-grained soil. In this book, coarse-grained soil is not regarded as a continuum material, but as a particulate material that is a multi-phased assembly, i.e., the soil consists of soil particles (solid phase) and pores filled with water (liquid phase) and/or air (gas phase). Furthermore, the macroscopic observed physical quantities, such as void ratio, water content, stress, strain and so on, are considered to be the mean values of the microscopic physical quantities, such as diameter of soil particle, contact angle of soil particles, number of contact points of soil particles and so on. These microscopic physical quantities are then regarded as random variables with various probability density functions.

Unsaturated Soil Mechanics with Probability and Statistics consists of the theoretical part of research works on unsaturated coarse-grained soil which has been performed for more than 30 years by Kitamura and for more than ten years by Kitamura and Sako. Computers are an essential tool for the proposed model which covers the seepage, water retention and deformation behaviors of saturated–unsaturated coarse-grained soil versatilely. Currently, computer science progresses rapidly and is applied to various fields and disciplines together with data assimilation for big data. The approach proposed in this book is considered to be available for the processing of observed macroscopic data obtained in the laboratory and in-situ tests, and field observations.

The authors wish that this book will be one of the textbooks that opens a new horizon for soil mechanics in the future, i.e., an epoch for the paradigm shift of discipline on soil mechanics.

Ryosuke KITAMURA

Kazunari SAKO

Acknowledgments

The research work in this book was started in 1974 when the first author Kitamura was a doctoral student supervised by the late Prof Murayama, Kyoto University. The late Prof Murayama often emphasized, in personal discussions with Kitamura, that original, intrinsic and versatile soil mechanics such as condensed matter physics had to be established in the field of the geotechnical engineering discipline. This book is considered to be one of the results of Prof Murayama's tutelage of Kitamura. The first author deeply acknowledges Prof Murayama, Kyoto University for his tutelage.

The authors greatly appreciate Mr Kazuyoshi Jomoto for his devoted contribution to drawing the perfect figures and tables. He has worked with Kitamura and Sako as the excellent technician of Kagoshima University for a long time.

Kitamura worked at the Department of Ocean Civil Engineering, Kagoshima University, Japan for 34 years from 1979 to 2013. More than 200 undergraduate and graduate students, including the co-author Sako, worked out of Kitamura's laboratory. Their excellent experimental and theoretical works encouraged the research work in this book greatly. We thank the students who graduated from Kitamura's laboratory. In particular, the authors wish to acknowledge the contribution from the following persons: Jorge Kisanuki, Makoto Takada, Yuji Miyamoto, Yoichi Uto, Kohei Araki and Mitsuhide Yamada.

Finally, the authors would like to give a lot of thanks to Mr Tony Moore, Ms Gabriella Williams and Mr Andrew Corrigan for their management of the editing of this book.

Ryosuke KITAMURA

Kazunari SAKO

Authors

Ryosuke Kitamura is an emeritus professor at Kagoshima University, Japan. He is a recipient of the 1983 Outstanding Paper Award for Young Researchers of the Japanese Geotechnical Society, the 2003 Outstanding Paper Award of Japan Society of Civil Engineers and JGS Medal for Merit for 2004, and has served as a chairman or member of several technical committees of JSCE and JGS.

Kazunari Sako is currently an associate professor of Kagoshima University, Japan. He is a recipient of the 2009 Outstanding Paper Award for Young Researchers of JGE. He serves as an editorial member of the Japanese Geotechnical Journal, *Soils and Foundations*, and the *Journal of the Japanese Society of Civil Engineering.*

Chapter 1

Introduction

1.1 BRIEF HISTORY OF MECHANICS LEADING TO PATH OF CURRENT SOIL MECHANICS

Figure 1.1 shows a brief history of the development of modern science from 17th century. Isaac Newton (1642–1727) published *Principia* (*Philosophiæ Naturalis Principia Mathematica*) in 1687, which was the start of the modern science era. The industrial revolution based on Newtonian mechanics (classical mechanics) developed in Europe (especially in the UK) from the 18th century and was followed by classical thermodynamics, electrodynamics, elastic mechanics and fluid mechanics. Classical thermodynamics led to statistical thermodynamics and then to the dynamics of complex systems. Electrodynamics developed into quantum mechanics, which is part of the mainstream of current physics. Elastic mechanics developed in the 18th–19th centuries, leading to elasto-plastic mechanics in the 20th century, then rational mechanics was established by assimilating with fluid mechanics in the 20th century. Rational mechanics is also called continuum mechanics and is part of the mainstream of current soil mechanics.

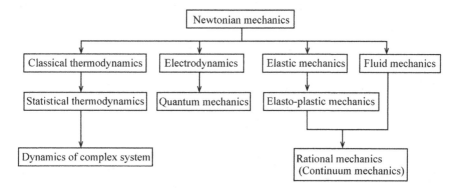

Figure 1.1 Brief history of modern science from 17th century.

Terzaghi published textbooks of soil mechanics in 1925 (in the German language) and 1943 (in the English language) which are theoretically based on elastic mechanics (Terghazi, 1925, 1943). These textbooks were the first specialized books for geotechnical engineering. Therefore he is now called the father of modern soil mechanics in the field of geotechnical engineering. Elastic mechanics is still available for current geotechnical engineering with respect to deformation ranges in small stress. Figure 1.2 shows a brief history of current soil mechanics, which has developed since 1925. In the 1960s, the Cambridge model (Schofield and Wroth, 1968) was proposed for saturated normally consolidated clay. Fredlund and Rahardjo (1993) and Fredlund et al. (2012) published textbooks of unsaturated soil mechanics in which their great works were summarized. These results and trends in the discipline of soil mechanics led to current theoretical soil mechanics. On the other hand, current theoretical soil mechanics is considered to be based on the continuum mechanics in Figure 1.1.

Figure 1.3 briefly shows the research history of mechanics for particulate material such as coarse-grained soil. Mogami (1967, 1969) tried to establish the intrinsic mechanics for soil and introduce the statistical thermodynamics approach to analyze the mechanical behaviors of coarse-grained soil. Murayama (1964, 1990) adopted the approach of condensed matter physics to analyze the mechanical behaviors of sand. The approach adopted in this book is considered to be further development of the basic concepts of Mogami and Murayama's research. Kitamura (1981a, 1981b) applied the probability theory (Markov process) to analyse the change in particulate soil structure of coarse-grained soil. Furthermore Sako and Kitamura (2006) proposed the mechanical model for the pore structure of coarse-grained soil to analyze the pore water retention and flow in unsaturated soil. Therefore this book is not based on continuum mechanics, but is

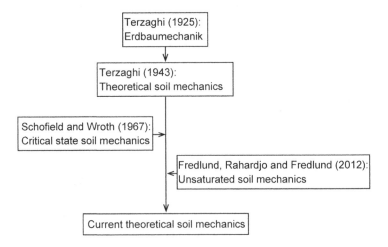

Figure 1.2 Brief history of modern theoretical soil mechanics.

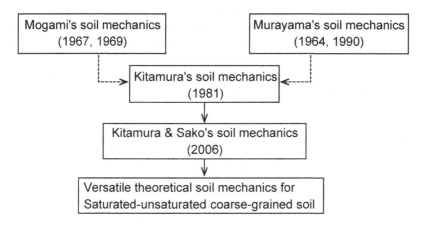

Figure 1.3 Brief history of mechanics for particulate material.

considered to be classified under the discipline of dynamics of complex systems in Figure 1.1.

1.2 SCOPE OF THIS BOOK

This book is a textbook on theoretical soil mechanics, but it is a little different from many other current books on soil mechanics, as this book aims to establish the mechanics not for continuum material, but for particulate material. Soil generally consists of soil particles (solid phase), pore water (fluid phase) and pore water (gas phase), i.e., it is a multi-phase material. The approach we adopted in this book starts from the standpoint that soil is a particulate material and an assembly of particles, water and air. Then the microscopic considerations are done to analyze the mechanical behaviors of soil particles, water and air on the scale of soil particles. Inevitably, probability theory and inferential statistics have to be introduced to estimate the macroscopic mechanical behaviors of soil from the microscopic consideration.

In this book, the microscopic mechanical model is proposed to analyze versatilely, the mechanical behaviors of saturated–unsaturated coarse-grained soil, which means that the authors aim to establish the versatile, not-so-sensitive mechanical model for coarse-grained soil using some basic physical quantities derived from some simple index soil tests, and applying probability theory and statistics to relate the microscopic physical quantities to the macroscopic ones. Here the coarse-grained soil is defined by the JGS (Japanese Geotechnical Society) as soil in which soil particles that are more than 0.075 mm in diameter occupy more than 50% in mass.

In Chapter 2, we review the probability theory and inferential statistics related to the proposed model for coarse-grained soil. First of all, the general hierarchic structure of population, sample populations and samples is explained. The attributes of the sample are also expressed in the sample space. Then the population, sample population and sample with the attributes are shown for the triaxial specimen of soil. The random variables and their probability density functions are introduced. The mean value and variance are defined using the probability density function. The Markov chain is also explained, because the Markov chain is applied to simulate the deformation behavior of coarse-grained soil in Chapter 9.

In Chapter 3, it is derived from the volume-mass equation for soil that the number of independent macroscopic physical quantities used to prescribe the state of soil is two, for example, the void ratio and water content. Then microscopic probabilistic models are respectively proposed for solid phase (particulate soil structure) and fluid phase (pore structure). The model for pore structure is called the elementary particulate model (EPM). Regarding two independent macroscopic physical quantities as the average of microscopic physical quantities, the microscopic physical quantities are related to two independent physical quantities. The void ratio and water content are adopted as two independent physical quantities. Additionally, the pore size distribution is also related to the grain size distribution, assuming that both are expressed by the logarithmic normal distributions.

In Chapter 4, the microscopic physical quantities, i.e., the number of particles per unit volume, the characteristic length, the numbers of contact points per unit volume and unit area, are derived by using the void ratio and the grain size distribution and applying the method to obtain the arithmetic average.

In Chapter 5, the inter-particle stress – defined as the average of inter-particle force per unit area – is introduced by using the method to obtain the arithmetic average. It is considered that the inter-particle stress is generated by the gravitational force, matric suction due to surface tension of pore water, seepage force due to flow of pore water and external force.

In Chapter 6, the soil-water characteristic curve is numerically derived by using the elementary particulate model (EPM) with the pore size distribution. The main drying, main wetting and hysteresis curves are numerically derived by using the ink-bottle model.

In Chapter 7, the unsaturated–saturated permeability coefficient with respect to the pore water is numerically derived by using the elementary particulate model (EPM) with pore size distribution. Additionally, the permeability coefficient with respect to the pore air is also derived.

In Chapter 8, the concept of average friction angle at contact points in soil block is introduced and the average friction angle is assumed to be same as the angle of repose of dry soil which is obtained from the inter-particle stress due to gravitational force. The potential slip plane is also defined by using the Mohr's stress circle. Then the apparent cohesion which depends on the suction and the self-weight retaining height are numerically derived. Finally, the methods to estimate the bearing capacity, earth pressure and slope stability are briefly introduced.

In Chapter 9, the microscopic model for deformation behavior is proposed. In the model, the motion of soil particles in soil block is microscopically considered to consist of the continuous motion, which is the change in contact angle at contact points and the discontinuous motion, which is the disappearance and appearance of contact points. The concept of potential slip plane is introduced to estimate the probabilistic change in contact angle with the change in stress state in the deformation process. Furthermore, the Markov chain is applied to estimate the change in the probability density function of contact angle with the change in stress state.

In Chapter 10, the microscopic physical quantities are calculated by using the grain size accumulation curve and void ratio obtained from the soil tests. These physical quantities are then used to calculate the interparticle stresses defined in Chapter 5. Furthermore, the soil-water characteristic curves with main drying curve, main wetting curve and scanning curves, the self-weight retaining height, the relationships between permeability coefficient and water content, and the stress–strain relation under the constant water content.

In Chapter 11, the assumptions introduced into the proposed model are summarized and then their issues to be solved are discussed in the conclusions.

REFERENCES

Fredlund, D. G. and Rahardjo, H. (1993). *Soil Mechanics for Unsaturated Soils.* John Wiley & Sons.

Fredlund, D. G., Rahardjo, H. and Fredlund, M. D. (2012). *Unsaturated Soil Mechanics in Engineering Practice.* John Wiley & Sons.

Kitamura, R., (1981a). A mechanical model of particulate material based on stochastic process. *Soils and Foundations*, 21(2), 64–72.

Kitamura, R., (1981b). Analysis of deformation mechanism of particulate material at particle scale. *Soils and Foundations*, 21(2), 85–98.

Mogami, T. (1967). Mechanics of granular material composed of particles of various sizes. *Trans. of JSCE*, No. 137, 42–47.

Mogami, T. (1969). *Mechanics for Particulate Material, Soil Mechanics Supervised by JSCE*, Gihodo Press, 893–1036 (in Japanese).

Murayama, S. (1964). A theoretical consideration on a behavior of sand. *Proc. IUTAM Symposium on Rheology and Soil Mechanics*, Grenoble, Springer-Verlag, 146–159.

Murayama, S. (1990). *Theory of Mechanical Behavior of Soils.* Gihodo Press (in Japanese).

Newton, I. (1687). *Philosophiæ Naturalis Principia Mathematica.*

Sako, K. and Kitamura, R. (2006). A practical numerical model for seepage behavior of unsaturated soil. *Soils and Foundations*, 46(5), 595–604.

Schofield, A. and Wroth, P. (1968). *Critical State Soil Mechanics.* McGraw-Hill Book Company.

Terzaghi, K. (1925). *Erdbaumechanik.* Franz Deuticke (in German).

Terzaghi, K. (1943). *Theoretical Soil Mechanics.* John Wiley & Sons.

Chapter 2

Review of probability theory and statistics

2.1 HIERARCHY OF POPULATION, SAMPLE POPULATION, AND SAMPLE

A universal set of samples can be described as a population which consists of samples in set theory. A group of samples taken from a population is called a sample population. A sample population can be sampled from a population which usually has a huge number of samples – more than 10^4. The sample population is applied to investigate the probabilistic properties of population in the inferential statistics. If the sample population is taken back to the population after sampling, the number of sampling is infinite (infinite population).

Figure 2.1 shows the conceptual diagram to explain the hierarchy of population, sample populations and samples. There are several methods to

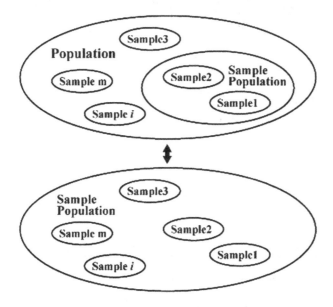

Figure 2.1 Population, sample population and samples.

use for sampling from the population to get a sample population, for example, random sampling, systematic sampling, etc. The probabilistic results obtained from the sample population are used to estimate the properties of the population in the inferential statistics by means of testing of hypotheses, confidence intervals, regression method, etc.

2.2 SAMPLE POINTS IN SAMPLE SPACE

A sample has its own attributes, for example, a human being has his (or her) own body height, body weight, etc., which are regarded as attributes of a sample. The attributes of samples can construct the sample points in the sample space. Figure 2.2 is a schematic diagram that shows the relation of the j-th sample population and points in a two-dimensional (2-D) sample space. When a human being is taken as a sample, x and y in Figure 2.2 correspond to the attributes of body height and body weight respectively and the sample point expresses the property of a human being in a 2-D sample space. A sample space is categorized as a discrete or continuous sample space, depending on the types of attributes.

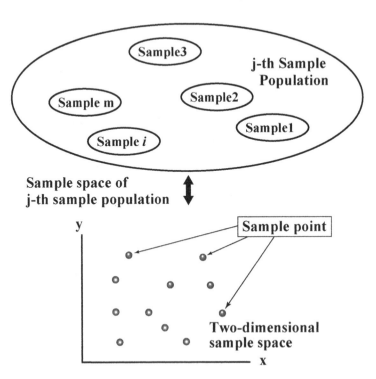

Figure 2.2 Samples and their sample points.

Let's consider a particulate material, such as coarse-grained soil which is called a multi-phase material consisting of a solid phase (soil particles), gas phase (pore air) and liquid (pore water) phase. The shape and size of soil particles are irregular and inevitably, the shape and size of pores filled with air and water will also vary. Therefore, trying to microscopically identify the state of particulate material and analyzing the motion and interaction of each phase might be the best way to introduce the probability theory and statistics.

Figure 2.3(a) shows a triaxial soil specimen of 5 cm in diameter and 10 cm in height. If the specimen consists of uniform spheres of 1 mm in diameter and the void ratio is 0.8, more than 2×10^5 particles are included in the specimen. This number is sufficient to treat the triaxial specimen as a population. Hence, it can be considered that an imaginary cube taken from the specimen by means of spoon, as shown in Figure 2.3(b), corresponds to a sample population. Furthermore, a cube with D_{cha} in length taken from the sample population shown in Figure 2.3(c) can be regarded as a sample, where D_{cha} is called the characteristic length; this will be discussed in Section 4.2.

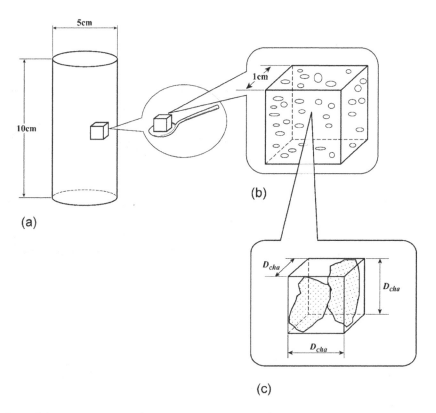

Figure 2.3 Population, sample population and sample for triaxial specimen. (a) Triaxial soil specimen (population), (b) Sample population, (c) Cube with D_{cha} in length (sample).

2.3 RANDOM VARIABLES AND PROBABILITY DISTRIBUTION

We can take a sample from the population as shown in Figure 2.1 and then obtain some quantitative (and/or qualitative) information from that sample. For example, the values of body height and body weight of a sample are attributes expressed by concrete values with respect to the quantitative information of a sample.

The qualitative information can be translated to figures and then treated as well as quantitative information.The variation of these values is random, depending on the sample taken out of the population. Therefore, we can define body height and body weight as random variables. Usually, the attribute of a random variable is expressed by a capital letter and the concrete value of a random variable is expressed by a lower-case letter as follows:

$$X = x \tag{2.3.1}$$

where
   X: random variable,
   x: concrete value of random variable X.

The event of $X=x$ is denoted by $\{X=x\}$ and the probability of the event $\{X=x\}$ is denoted by $P\{X=x\}$. The random variable is denoted by the capital letter and its concrete value is denoted by the small letter in the following.

The properties of random variables can be estimated by the probability distribution which can be expressed by the probability mass function (PMF) for the discrete sample space, and the probability density function (PDF) for the continuous sample space, respectively. The probability distribution of random variable X for the continuous sample space is schematically shown in Figure 2.4, where the properties of random variable X can be expressed by both the PDF $f(x)$ and the cumulative distribution function (CDF) $F(x)$ for continuous random variables. The following conditions should be satisfied for the PDF $f(x)$:

$$f(x) \geq 0 \tag{2.3.2}$$

$$\int_{-\infty}^{\infty} f(x)dx = 1 \tag{2.3.3}$$

The probability of which the random variable X ranged between a and b, $P\{a < X \leq b\}$ can be calculated by the following equation for continuous sample space:

$$P\{a < X \leq b\} = \int_{a}^{b} f(x)dx = F(b) - F(a) \tag{2.3.4}$$

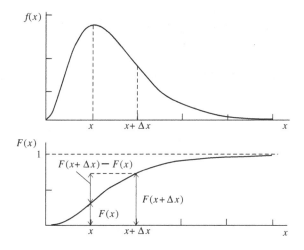

Figure 2.4 Relation between probability density function (PDF) *f(x)* and cumulative distribution function (CDF) *F(x)*.

Furthermore, the probability of a random variable ranged between x and $x + \Delta x$, $P\{x < X \le x + \Delta x\}$ can be expressed as follows:

$$P\{x < X \le x + \Delta x\} \cong f\left(x + \frac{\Delta x}{2}\right) \cdot \Delta x = F(x + \Delta x) - F(x) \qquad (2.3.5)$$

The relation between the PDF $f(x)$ and the CDF $F(x)$ can be expressed by the following equations:

$$F(x) = P\{X \le x\} = \int_{-\infty}^{x} f(x) dx \qquad (2.3.6)$$

or

$$f(x) = \frac{dF(x)}{dx} \qquad (2.3.7)$$

2.4 PARAMETERS OF PROBABILITY DISTRIBUTION

2.4.1 Mean value and variance

Let's consider the sample population which includes N samples. When the concrete value of attributes of the i-th sample in the population is denoted as x_i in the one-dimensional sample space, the mean value of the random variable X_i can be obtained by the following equation for the discrete sample space:

$$\mu = \bar{x} = E[X_i] = \frac{1}{N} \sum_{i=1}^{N} x_i \qquad (2.4.1)$$

where

μ, $\bar{x}$, $E[X_i]$: mean value of random variable X_i and

N: number of samples in the sample population.

Equation (2.4.1) is written as follows:

$$\sum_{i=1}^{N} x_i = N \cdot \bar{x} = N \cdot E[X_i] \qquad (2.4.2)$$

Introducing the histogram (frequency distribution), the mean value is obtained by the following equation in the discontinuous sample space:

$$\bar{x} = \frac{1}{N} \sum_{i=1}^{n} N_i \cdot x_i = \sum_{i=1}^{n} \frac{N_i}{N} \cdot x_i = \sum_{i=1}^{n} p(x_i) \cdot x_i \qquad (2.4.3)$$

where

N_i: number of samples with sampled value of x_i,

n: number of classes in the histogram,

$$N = \sum_{i=1}^{n} N_i$$

and $p(x_i)$: relative frequency (PMF).

Then N_i is obtained from Equation (2.4.3) as follows:

$$N_i = N \cdot p(x_i) \qquad (2.4.4)$$

The variance of random variable X_i in the discrete sample space is obtained from the following equation:

$$\sigma^2 = V[X_i] = \sum_{i=1}^{N} (x_i - \bar{x})^2 p(x_i)$$

$$= \sum_{i=1}^{N} x_i^2 p(x_i) - 2\bar{x} \sum_{i=1}^{N} x_i p(x_i) + \bar{x}^2 \sum_{i=1}^{N} p(x_i) \qquad (2.4.5)$$

$$= E[X_i^2] - 2\bar{x}^2 + \bar{x}^2 = E[X_i^2] - \bar{x}^2$$

where σ^2, $V[X_i]$: variance of random variable X_i.

Equation (2.4.5) is rewritten as follows:

$$V[X_i] = E[X_i^2] - \mu^2 \qquad (2.4.6)$$

For the continuous sample space, Equations, (2.4.1) and (2.4.4) are expressed as follows:

$$\mu = \bar{x} = E[X] = \int_{-\infty}^{\infty} x \cdot f(x) dx \qquad (2.4.7)$$

where $f(x)$: PDF of random variable X.

$$N_i = N \cdot f(x) \cdot dx \qquad (2.4.8)$$

The variance of random variable X is expressed for the continuous sample space, corresponding to Equation (2.4.5) as follows:

$$\sigma^2 = V[X] = \int_{-\infty}^{\infty} (x - \bar{x})^2 f(x) dx$$

$$= \int_{-\infty}^{\infty} x^2 \cdot f(x) dx - 2\bar{x} \int_{-\infty}^{\infty} x \cdot f(x) dx + \bar{x}^2 \int_{-\infty}^{\infty} f(x) dx \qquad (2.4.9)$$

$$= E[X^2] - \bar{x}^2$$

Equation (2.4.9) is rewritten as follows:

$$V[X] = E[X^2] - \mu^2 \qquad (2.4.10)$$

When a sample has 2 attributes designated by x and y and these the random variables X and Y are independent, the joint PDF $f(x,y)$ is rewritten as follows:

$$f(x,y) = f_x(x) \cdot f_y(y) \qquad (2.4.11)$$

where
    $f(x,y)$: joint PDF of random variables X and Y,
    $f_x(x)$: marginal probability function of random variable X,
    $f_y(y)$: marginal probability function of random variable Y.

Introducing the function of random variables X and Y, $\varphi(X, Y)$, the mean value of $\varphi(X, Y)$ is obtained as follows:

$$E[\varphi(X,Y)] = \iint \varphi(x,y) \cdot f(x,y) dx dy \qquad (2.4.12)$$

When the random variables X and Y are independent, Equation (2.4.12) is rewritten as follows:

$$E[\varphi(X,Y)] = \iint \varphi(x,y) \cdot f_x(x) \cdot f_y(y) dx dy \qquad (2.4.13)$$

2.4.2 Coefficient of variation

The coefficient of variation κ is defined as the ratio of standard deviation σ to the mean value μ as follows:

$$\kappa = \frac{\sigma}{\mu} \qquad (2.4.14)$$

It is found from Equation (2.4.13) that the coefficient of variation corresponds to the standard deviation with the unit mean value.

2.5 NORMAL DISTRIBUTION AND LOGARITHMIC NORMAL DISTRIBUTION

2.5.1 Normal distribution

The normal distribution $f(x)$ of random variable X is expressed by the following equation, and its curve is shown in Figure 2.5, where the inflection points are located at $x = \pm\sigma$:

$$f(x) = \frac{1}{\sqrt{2\pi}\sigma} \exp\left[-\frac{(x-\mu)^2}{2\sigma^2}\right] \qquad (2.5.1)$$

where
   μ: mean (or expected) value,
   σ: standard deviation,
   σ^2:   variance and
   x: concrete value of random variable X.

The normal distribution with the mean value μ and standard deviation σ is denoted by $N(\mu,\sigma)$. It is found from Equation (2.5.1) that normal distribution is prescribed by two distribution parameters, μ and σ.

Instead of random variable X, we introduce another random variable U, defined as follows:

$$u = \frac{x-\mu}{\sigma} \qquad (2.5.2)$$

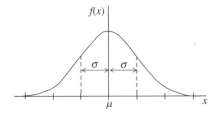

Figure 2.5 Normal distribution.

Then the differential of Equation (2.5.2) is obtained as follows:

$$dx = \sigma \cdot du \tag{2.5.3}$$

Substituting Equations (2.5.2) and (2.5.3) into Equation (2.5.1), the following equation is obtained:

$$f(x)\,dx = \frac{1}{\sqrt{2\pi}\sigma}\exp\left[-\frac{(x-\mu)^2}{2\sigma^2}\right]dx$$

$$= \frac{1}{\sqrt{2\pi}}\exp\left[-\frac{u^2}{2}\right]du = \phi(u)\,du \tag{2.5.4}$$

where

$$\phi(u) = \frac{1}{\sqrt{2\pi}}\exp\left[-\frac{u^2}{2}\right] \tag{2.5.5}$$

$\phi(u)$ in Equation (2.5.5) is called the standard normal distribution of which the mean value and standard deviation are 0 and 1 respectively, i.e., N $(0,1)$. The standard normal distribution $\phi(u)$ is tabulated as tables of normal probabilities and is widely used to obtain the value of probability for the random variable U which is ranged between $u = 0$ and other concrete value $u = u_\alpha > 0$.

2.5.2 Logarithmic normal distribution

If the random variable X in Equation (2.5.1) is replaced by $\ln X$, the following equation can be obtained:

$$f(\ln x) = \frac{1}{\sqrt{2\pi}\varsigma}\exp\left[-\frac{(\ln x - \lambda)^2}{2\varsigma^2}\right] \tag{2.5.6}$$

where
    λ: mean value of $\ln x$,
    ς: standard deviation of $\ln x$.

The following equation is introduced to transform the variable x into the new variable y:

$$y = \ln x \tag{2.5.7}$$

Then Equation (2.5.7) is differentiated as follows:

$$dy = \frac{1}{x}dx \tag{2.5.8}$$

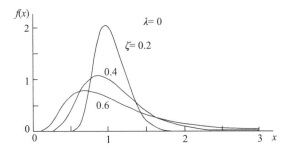

Figure 2.6 Logarithmic normal distribution on the normal scale with $\lambda = 0$ and various values of ζ.

Equation (2.5.6) can be reduced to the following equation similar to Equation (2.5.4), using Equations (2.5.7) and (2.5.8):

$$f(\ln x)d(\ln x) = f(y)dy = f(y)\cdot\frac{1}{x}\cdot dx$$

$$= \frac{1}{\sqrt{2\pi}\cdot\varsigma}\exp\left[-\frac{(y-\lambda)^2}{2\varsigma^2}\right]\cdot\frac{1}{x}\cdot dx$$

$$= \frac{1}{\sqrt{2\pi}\cdot\varsigma\cdot x}\exp\left[-\frac{(\ln x-\lambda)^2}{2\varsigma^2}\right]dx \qquad (2.5.9)$$

$$= f_{\ln}(x)dx$$

where

$$f_{\ln}(x) = \frac{1}{\sqrt{2\pi}\cdot\varsigma\cdot x}\exp\left[-\frac{(\ln x-\lambda)^2}{2\varsigma^2}\right] \qquad (2.5.10)$$

The logarithmic normal distribution can be expressed using both Equations (2.5.6) and (2.5.10), which are on the logarithmic scale and linear scale respectively. Figure 2.6 shows the logarithmic normal distribution expressed by Equation (2.5.10) with $\lambda=0$ and various values of ζ. Note that the scale of the horizontal axis in Figure 2.6 is not logarithmic, but linear.

2.5.3 Relations between mean values and variances of logarithmic normal distribution expressed by linear scale and logarithmic scale

The mean value of logarithmic normal distribution on the linear scale can be obtained by using Equations (2.4.7) and (2.5.10) as follows:

$$\mu = E[X] = \int_0^\infty x \cdot f_{\ln}(x)\,dx$$

$$= \int_0^\infty x \cdot \frac{1}{\sqrt{2\pi}\varsigma \cdot x} \exp\left[-\frac{(\ln x - \lambda)^2}{2\varsigma^2}\right]dx \tag{2.5.11}$$

Using Equations (2.5.7) and (2.5.8), the following equation can be obtained:

$$dx = x \cdot dy = e^y dy \tag{2.5.12}$$

Substituting Equation (2.5.12) into Equation (2.5.11), the following equation is obtained:

$$\mu = \int_0^\infty x \cdot f_{\ln}(x)\,dx$$

$$= \int_{-\infty}^\infty \frac{1}{\sqrt{2\pi}\varsigma} \cdot e^y \cdot \exp\left[-\frac{(y-\lambda)^2}{2\varsigma^2}\right]dy \tag{2.5.13}$$

$$= \int_{-\infty}^\infty \frac{1}{\sqrt{2\pi}\varsigma} \cdot \exp\left[y - \frac{(y-\lambda)^2}{2\varsigma^2}\right]dy$$

The exponential part in Equation (2.5.13) can be reduced as follows:

$$y - \frac{(y-\lambda)^2}{2\varsigma^2} = -\frac{1}{2\varsigma^2}\left\{-2\varsigma^2 y + (y-\lambda)^2\right\}$$

$$= -\frac{1}{2\varsigma^2}\left\{y^2 - 2(\lambda+\varsigma^2)y + \lambda^2\right\}$$

$$= -\frac{1}{2\varsigma^2}\left[\left\{y-(\lambda+\varsigma^2)\right\}^2 - (2\lambda\varsigma^2 + \varsigma^4)\right] \tag{2.5.14}$$

$$= -\frac{1}{2\varsigma^2}\left[\left\{y-(\lambda+\varsigma^2)\right\}^2\right] + \left(\lambda+\frac{\varsigma^2}{2}\right)$$

Substituting Equation (2.5.14) into Equation (2.5.13), the following equation is obtained:

$$\mu = \int_{-\infty}^\infty \frac{1}{\sqrt{2\pi}\varsigma} \cdot \exp\left[-\frac{1}{2\varsigma^2}\left[\left\{y-(\lambda+\varsigma^2)\right\}^2\right] + \left(\lambda+\frac{\varsigma^2}{2}\right)\right]dy$$

$$= \exp\left(\lambda+\frac{\varsigma^2}{2}\right) \cdot \int_{-\infty}^\infty \frac{1}{\sqrt{2\pi}\varsigma} \cdot \exp\left[-\frac{1}{2\varsigma^2}\left[\left\{y-(\lambda+\varsigma^2)\right\}^2\right]\right]dy \tag{2.5.15}$$

$$= \exp\left(\lambda+\frac{\varsigma^2}{2}\right) \cdot \int_{-\infty}^\infty \frac{1}{\sqrt{2\pi}\varsigma} \cdot \exp\left[-\frac{\left\{y-(\lambda+\varsigma^2)\right\}^2}{2\varsigma^2}\right]dy$$

In Equation (2.5.15), the part of integration becomes:

$$\int_{-\infty}^{\infty} \frac{1}{\sqrt{2\pi}\varsigma} \cdot \exp\left[-\frac{\left\{y-\left(\lambda+\varsigma^2\right)\right\}^2}{2\varsigma^2}\right]dy = 1 \tag{2.5.16}$$

Hence Equation (2.5.15) is rewritten as:

$$\mu = \exp\left(\lambda+\frac{\varsigma^2}{2}\right) \tag{2.5.17}$$

Equation (2.5.17) can be rewritten as:

$$\lambda = \ln\mu - \frac{\varsigma^2}{2} \tag{2.5.18}$$

Similarly, the mean value of random variable X^2 can be obtained as follows:

$$E\left[X^2\right] = \int_0^{\infty} x^2 \cdot f_{\ln}(x)dx$$

$$= \int_0^{\infty} x^2 \cdot \frac{1}{\sqrt{2\pi}\cdot\varsigma\cdot x}exp\left[-\frac{(lnx-\lambda)^2}{2\varsigma^2}\right]dx$$

$$= \int_{-\infty}^{\infty} \frac{1}{\sqrt{2\pi}\varsigma} \cdot e^{2y} \cdot \exp\left[-\frac{(y-\lambda)^2}{2\varsigma^2}\right]dy \tag{2.5.19}$$

$$= \int_{-\infty}^{\infty} \frac{1}{\sqrt{2\pi}\varsigma} \cdot \exp\left[-\frac{\left\{y^2-2\left(\lambda+2\varsigma^2\right)y+\lambda^2\right\}}{2\varsigma^2}\right]dy$$

$$= \exp\left\{2\left(\lambda+\varsigma^2\right)\right\}\int_{-\infty}^{\infty} \frac{1}{\sqrt{2\pi}\varsigma} \cdot \exp\left[-\frac{\left\{y-\left(\lambda+2\varsigma^2\right)\right\}^2}{2\varsigma^2}\right]dy$$

In Equation (2.5.19), the part of integration become:

$$\int_{-\infty}^{\infty} \frac{1}{\sqrt{2\pi}\varsigma} \cdot \exp\left[-\frac{\left\{y-\left(\lambda+2\varsigma^2\right)\right\}^2}{2\varsigma^2}\right]dy = 1 \tag{2.5.20}$$

Hence, Equation (2.5.19) is rewritten as follows:

$$E\left[X^2\right] = \exp\left\{2\left(\lambda + \varsigma^2\right)\right\} = \exp\left\{2\left(\lambda + \frac{\varsigma^2}{2}\right) + \varsigma^2\right\}$$

$$= \mu^2 \cdot \exp\left(\varsigma^2\right) \tag{2.5.21}$$

Substituting Equation (2.5.21) into Equation (2.4.6), the following equation is obtained, using Equation (2.4.6):

$$\sigma^2 = V\left[X\right] = E\left[X^2\right] - \mu^2 = \mu^2\left(e^{\varsigma^2} - 1\right) \tag{2.5.22}$$

The following equation is then obtained from Equation (2.5.22):

$$\varsigma^2 = \ln\left(1 + \frac{\sigma^2}{\mu^2}\right) \tag{2.5.23}$$

Finally, the logarithmic distribution parameters (λ, ζ) on the logarithmic scale are related to (μ, σ) on the linear scale and by Equations (2.5.18) and (2.5.23). The logarithmic normal distribution is also prescribed by two distribution parameters, (λ, ζ) or (μ, σ), as well as the normal distribution.

Figure 2.7 shows two CDFs of logarithmic normal distributions, CDF1 and CDF2, which have the same coefficient of variation $\kappa = \sigma/\mu$ and different mean values. The following equation is then obtained:

$$\kappa = \frac{\sigma_1}{\mu_1} = \frac{\sigma_2}{\mu_2} \tag{2.5.24}$$

where
μ_1 and μ_2: mean values of CDF1 and CDF2 on the linear scale and
σ_1 and σ_2: variances of CDF1 and CDF2 on the linear scale.

Substituting Equation (2.5.24) into Equation (2.5.23), the following equation is obtained:

$$\zeta_1 = \zeta_2 \tag{2.5.25}$$

where ζ_1, ζ_2: variances of CDF1 and CDF2 on the logarithmic scale.

From Equation (2.5.25), we find that two CDFs, CDF1 and CDF2, expressed by the logarithmic scale are similar in shape and match each other with parallel movement, as shown in Figure 2.7.

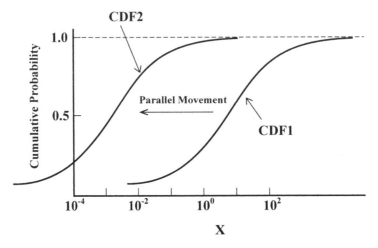

Figure 2.7 Two CDFs, CDFI and CDF2, with same coefficient of variation and different mean values.

2.6 REGRESSION ANALYSIS

2.6.1 Linear regression analysis

Figure 2.8 shows n data plots of two variables (x_i, y_i), which are obtained by sampling. It is then assumed that the relation between x and y can be expressed by the following linear equation:

$$y_i = ax_i + b + e_i \quad (i = 1 \sim n) \tag{2.6.1}$$

where

x_i: explanatory variable,
y_i: objective variable,
e_i: error and
a, b: constants.

The summation of squares of errors is obtained as follows, using Equation (2.6.1).

$$S = \sum_{i=1}^{n} e_i^2 = \sum_{i=1}^{n} \left\{ y_i - (ax_i + b) \right\}^2 \tag{2.6.2}$$

The coefficients of regression can be obtained by minimizing Equation (2.6.2), i.e., the normal equations are expressed as follows.

$$\frac{\partial S}{\partial a} = -2 \sum_{i=1}^{n} x_i \left\{ y_i - (ax_i + b) \right\} = 0 \tag{2.6.3}$$

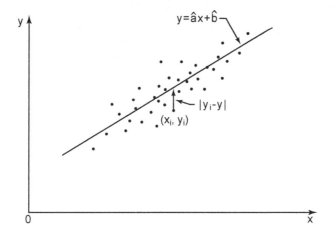

Figure 2.8 Data of two variables (x_i, y_i) and optimum relation by linear function.

$$\frac{\partial S}{\partial b} = -2\sum_{i=1}^{n} x_i \left\{ y_i - (ax_i + b) \right\} = 0 \tag{2.6.4}$$

Solving simultaneous equations, Equations (2.6.3) and (2.6.4), the following equations are obtained, although the reduction is omitted:

$$\hat{a} = \frac{\displaystyle\sum_{i=1}^{n} (x_i - \bar{x})(y_i - \bar{y})}{\displaystyle\sum_{i=1}^{n} (x_i - \bar{x})^2} \tag{2.6.5}$$

$$\hat{b} = \bar{y} - a\bar{x} \tag{2.6.6}$$

where $\hat{a}$ and $\hat{b}$: coefficients of regression (estimated value of a and b in Equation (2.6.1)),

$$\bar{x} = \frac{1}{n}\sum_{i=1}^{n} x_i \tag{2.6.7}$$

$$\bar{y} = \frac{1}{n}\sum_{i=1}^{n} y_i \tag{2.6.8}$$

Consequently, the regression equation presenting the relation between x and y in Figure 2.8 is expressed as follows:

$$y = \hat{a}x + \hat{b} \tag{2.6.9}$$

2.6.2 Non-linear regression analysis

2.6.2.1 Normal distribution

Let's consider fitting n data plots of two variables (x_i, y_i) to normal distribution where the variable y_i is defined as follows:

$$y_i = F(x_i) = \int_{-\infty}^{x_i} f(x)\,dx = \int_{-\infty}^{u_i} \phi(u)\,du = \Phi(u_i) \tag{2.6.10}$$

where

$\phi(u)$: standard normal distribution,

$\Phi(u_i)$: CDF of $\phi(u)$ given by numerical table.

$$f(x) = \frac{1}{\sqrt{2\pi}\sigma} \exp\left[-\frac{(x-\mu)^2}{2\sigma^2}\right] \tag{2.5.1bis}$$

$$\phi(u) = \frac{1}{\sqrt{2\pi}} \exp\left[-\frac{u^2}{2}\right] \tag{2.5.5bis}$$

The following equation is obtained from Equation (2.6.10):

$$u_i = \Phi^{-1}(y_i) \tag{2.6.11}$$

The following equation is then introduced for n data plots of two variables (x_i, y_i), using Equation (2.5.2).

$$x_i = \sigma u_i + \mu + e_i \tag{2.6.12}$$

where

μ: mean value of normal distribution,

σ: standard deviation of normal distribution and

e_i: error.

Equation (2.6.12) corresponds to Equation (2.6.1) and the following equation is then obtained with respect to the summation of squares of error:

$$S = \sum_{i=1}^{n} e_i^2 = \sum_{i=1}^{n} \left\{x_i - (\sigma u_i + \mu)\right\}^2 \tag{2.6.13}$$

Finally, the estimated mean value $\hat{\mu}$ and standard deviation $\hat{\sigma}$ are obtained as follows, although the reduction is omitted:

$$\hat{\sigma} = \frac{\sum_{i=1}^{n} (x_i - \bar{x}) \cdot (u_i - \bar{u})}{\sum_{i=1}^{n} (u_i - \bar{u})^2} \tag{2.6.14}$$

$$\hat{\mu} = \bar{x} - \hat{\sigma} \cdot \bar{u} \tag{2.6.15}$$

where $\hat{\sigma}$ and $\hat{\mu}$: coefficients of regression (estimated value of σ and μ in Equation (2.6.12)),

$$\bar{x} = \frac{1}{n} \sum_{i=1}^{n} x_i \tag{2.6.7bis}$$

$$\bar{u} = \frac{1}{n} \sum_{i=1}^{n} u_i \tag{2.6.8bis}$$

2.6.2.2 Logarithmic normal distribution

Let's consider fitting n data plots of two variables (x_i, y_i) to logarithmic normal distribution. The logarithmic normal distribution is expressed as follows:

$$f(\ln x) = \frac{1}{\sqrt{2\pi}\varsigma} \exp\left[-\frac{(\ln x - \lambda)^2}{2\varsigma^2} \right] \tag{2.5.6bis}$$

The following equation is introduced to transform the variable $\ln x$ into r:

$$\frac{\ln x - \lambda}{\varsigma} = r \tag{2.6.16}$$

The following equation is obtained by differentiating both sides of Equation (2.6.16):

$$dx = \varsigma \cdot x \cdot dr \tag{2.6.17}$$

Substituting Equations (2.6.16) and (2.6.17) into Equation (2.5.6), the following equation is obtained:

$$y_i = F_{\ln}(x_i) = \int_{-\infty}^{r_i} \frac{1}{\sqrt{2\pi}} \cdot \exp\left(-\frac{r^2}{2} \right) dr = \Phi(r_i) \tag{2.6.18}$$

where $\Phi(r_i)$: CDF of standard normal distribution.
    The following equation is then obtained:

$$r_i = \Phi^{-1}(y_i) \tag{2.6.19}$$

Introducing another variable, z defined by $z = \ln x$, the following equation is obtained from Equation (2.6.16) in a similar way to regression analysis for normal distribution:

$$z_i = \ln x_i = \varsigma r_i + \lambda + e_i \tag{2.6.20}$$

where

λ: mean value of logarithmic normal distribution,

ζ: standard deviation of logarithmic normal distribution and

e_i: error.

Equation (2.6.20) corresponds to Equations (2.6.1) and (2.6.12), and then the following equation is obtained with respect to the summation of squares of error:

$$S = \sum_{i=1}^{n} e_i^2 = \sum_{i=1}^{n} \left\{ z_i - \left(\zeta r_i + \lambda \right) \right\}^2 \tag{2.6.21}$$

Finally, the estimated mean value $\hat{\lambda}$ and standard deviation $\hat{\zeta}$ are obtained as follows, although the reduction is omitted:

$$\hat{\zeta} = \frac{\displaystyle\sum_{i=1}^{n} \left(z_i - \bar{z} \right) \cdot \left(r_i - \bar{r} \right)}{\displaystyle\sum_{i=1}^{n} \left(r_i - \bar{r} \right)^2} \tag{2.6.22}$$

$$\hat{\lambda} = \bar{z} - \hat{\zeta} \cdot \bar{r} \tag{2.6.23}$$

where $\hat{\zeta}$ and $\hat{\lambda}$: coefficients of regression (estimated value of ζ and λ in Equation (2.6.20)),

$$\bar{z} = \frac{1}{n} \sum_{i=1}^{n} z_i \tag{2.6.24}$$

$$\bar{r} = \frac{1}{n} \sum_{i=1}^{n} r_i \tag{2.6.25}$$

2.7 MARKOV CHAIN

The random walking model is well-known as one of the Markov chain. The Markov chain can be mathematically expressed by the following equation:

$$P\left\{ A_j \left(k+1 \right) \right\} = \sum_{i=1}^{r} P\left\{ A_i \left(k \right) \right\} \cdot P\left\{ A_j \left(k+1 \right) \, \middle| \, A_i \left(k \right) \right\} \tag{2.7.1}$$

where

r: number of events,

$P\left\{ A_i \left(k \right) \right\}$: probability of $\left\{ A_i \left(k \right) \right\}$ that the event A_i comes out at the time k,

$P\{A_j(k+1)\}$:   probability of $\{A_j(k+1)\}$ that the event A_j comes

out at the time $(k+1)$ and

$P\{A_j(k+1)\,|A_i(k)\}$: conditional probability of the event A_j at the time

$(k+1)$ under the condition that the event A_i comes

out at the time k.

The conditional probability $P\{A_j(k+1)\,|A_i(k)\}$ can be abbreviated as follows:

$$p_{ij} = P\{A_j(k+1)\,|A_i(k)\} \qquad (2.7.2)$$

where p_{ij}: transition probability that the state i at the time k changes to the state j at the time $k+1$.

Equation (2.7.1) is then rewritten as follows:

$$P\{A_j(k+1)\} = \sum_{i=1}^{r} p_{ij} \cdot P\{A_i(k)\} \qquad (2.7.3)$$

Furthermore, when the probability $P\{A_i(k)\}$ is expressed by $p_i(k)$, Equation (2.7.3) is rewritten as follows:

$$p_j(k+1) = \sum_{i=1}^{r} p_{ij} \cdot p_i(k) \qquad (2.7.4)$$

Figure 2.9 shows the schematic diagram to relate the probability of state j at the time k + 1, $p_j(k+1)$, to the probability of states i $(i=1\sim r)$ at the time k, $p_i(k)$ by the transition probability p_{ij} $(i=1\sim r)$.

Furthermore, the probability $p_j(k+2)$ is expressed as follows, using Equation (2.7.4):

$$p_j(k+2) = \sum_{l=1}^{r} p_{lj} \cdot p_l(k+1) = \sum_{l=1}^{r} p_{lj} \cdot \sum_{i=1}^{r}(p_{il} \cdot p_i(k))$$

$$= \sum_{i=1}^{r}\left(\sum_{l=1}^{r} p_{il} \cdot p_{lj}\right) \cdot p_i(k) = \sum_{i=1}^{r} p_{ij}^{(2)} \cdot p_i(k) \qquad (2.7.5)$$

where

$$p_{ij}^{(2)} = \sum_{l=1}^{r} p_{il} \cdot p_{lj} : 2\text{-step transition probability} \qquad (2.7.6)$$

$p_{ij}^{(2)}$ is extended to n-step transition probability as follows:

$$p_{ij}^{(n)} = \sum_{l=1}^{r} p_{il} \cdot p_{lj}^{(n-1)} = \sum_{l=1}^{r} p_{il}^{(n-m)} \cdot p_{lj}^{(m)} \qquad (2.7.7)$$

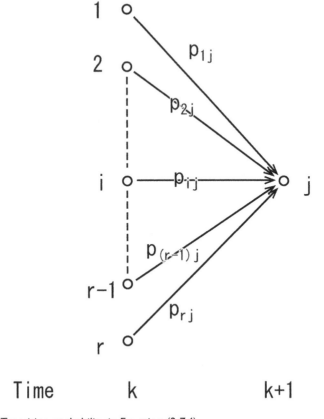

Figure 2.9 Transition probability in Equation (2.7.4).

Finally, the Markov chain is generally expressed by the following equation:

$$p_j(n) = \sum_{i=1}^{r} p_{ij}^{(n)} \cdot p_i(0) \tag{2.7.8}$$

where

$$\sum_{j=1}^{r} p_j(n) = 1 \tag{2.7.9}$$

The Markov chain will be applied microscopically to analyze the change in particulate soil structure which is discussed in Chapter 9.

Chapter 3

Microscopic models of soil using probability distributions

3.1 MACROSCOPIC PHYSICAL QUANTITIES OF SATURATED–UNSATURATED SOIL AND THEIR PHASE DIAGRAM

Figure 3.1 shows the microscopic state of the volcanic ash called "Shirasu" in Kagoshima Prefecture, Japan; this is from a photo taken by Haruyama (1969) using an optical microscope. "Shirasu" is the local name for the coarse-grained soil and is geologically defined as the non-welded part of pyroclastic flow deposits. Shirasu covers the ground surface in Kagoshima Prefecture widely. The Shirasu particles shown in Figure 3.1 are estimated to be derived from the enormous eruption of Aira Caldera about 30,000 years ago.

It can be generally extended from Figure 3.1 that the microscopic state of unsaturated coarse-grained soil can be schematically drawn as shown in Figure 3.2. It is found from Figures 3.1 and 3.2 that soils generally consist of soil particles (solid phase) with pores (water and gas phases) among them, i.e., it is a multi-phase material. In conventional soil mechanics, these phases in soil blocks are separated and summed up to draw the phase diagram shown in Figure 3.3(a) in which the left and right sides denote the mass and volume of a soil block respectively. Separating and summing up the mass and volume of three phases is considered to be a modeling techniques for a soil block. A little air in the water phase and a little water in the air phase are contained in real soil according to Henry's law, but in Figure 3.3(a) these are not taken account of. Figure 3.3(b) shows the alternative phase diagram where the mass and volume of soil particles are respectively the units, and Figure 3.3(c) shows another phase diagram where the volume of soil particles is the unit. The air density is about 1/1000 of water density and then the mass of air is neglected in Figure 3.3. The phase diagram shown in Figure 3.3 can be regarded as one of the models of soil used to identify the state of unsaturated soil.

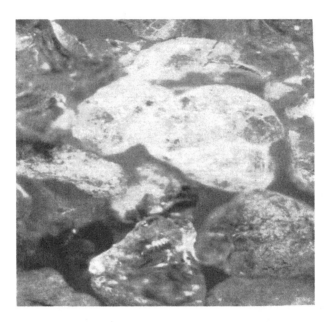

Figure 3.1 Volcanic ash with size of 0.42–0.84 mm taken by optical microscope (after Haruyama, 1969).

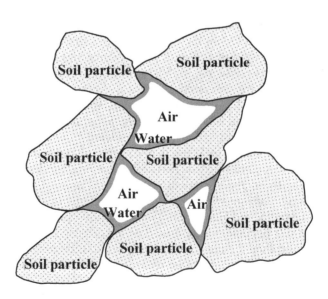

Figure 3.2 Schematic state of unsaturated coarse-grained soil on a microscopic scale.

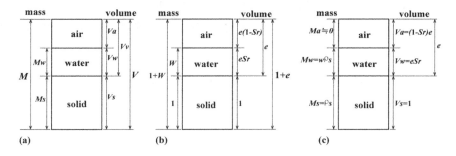

Figure 3.3 Phase diagram of soil (a) Phase diagram, (b) Phase diagram for unit mass and unit volume of solid phase, (c) Phase diagram for unit volume of solid phase.

The macroscopic physical quantities of soil are then defined as follows, referring to Figure 3.3(a), (b) and (c):

Soil particle density: $\rho_s = \dfrac{M_s}{V_s}$ (3.1.1)

Void ratio: $e = \dfrac{V_v}{V_s}$ (3.1.2)

Water content: $w = \dfrac{M_w}{M_s}$ (3.1.3)

Porosity: $n = \dfrac{V_v}{V} = \dfrac{e}{1+e}$ (3.1.4)

where $(1+e)$: volume of a soil block when the volume of the solid part is the unit shown in Figure 3.3(b), which is called the specific volume in conventional soil mechanics.

In Figure 3.3(c), the following equation is derived for the liquid phase:

$$w \cdot \rho_s = \rho_w \cdot e \cdot S_r \qquad\qquad\qquad\qquad (3.1.5)$$

where ρ_w: water density.

Equation (3.1.5) is rewritten as follows:

$$e \cdot S_r = w \cdot \frac{\rho_s}{\rho_w} = w \cdot G_s \qquad\qquad\qquad (3.1.6)$$

where G_s: Specific gravity of soil particle.

The degree of saturation S_r is then obtained as follows:

$$S_r = \frac{V_w}{V_v} = \frac{w \cdot G_s}{e} \tag{3.1.7}$$

The water porosity (volumetric water content) n_w and the air porosity (effective porosity) n_a are respectively obtained as follows:

$$n_w = \frac{V_w}{V} = \frac{w \cdot G_s}{1+e} \tag{3.1.8}$$

$$n_a = \frac{V_a}{V} = \frac{e - w \cdot G_s}{1+e} \tag{3.1.9}$$

Furthermore, the dry density ρ_d and wet density ρ_t are respectively obtained as follows:

$$\rho_d = \frac{M_s}{V} = \frac{\rho_s}{1+e} \tag{3.1.10}$$

$$\rho_t = \frac{M}{V} = \frac{M_s + M_w}{V} = \frac{1+w}{1+e} \cdot \rho_s \tag{3.1.11}$$

As the degree of saturation S_r is 1 for saturated soil, Equation (3.1.6) is rewritten as follows:

$$e = w \cdot G_s \tag{3.1.12}$$

The saturated density ρ_{sat} is then obtained as follows, using Equations (3.1.11) and (3.1.12):

$$\rho_{sat} = \frac{1 + e/G_s}{1+e} \cdot \rho_s = \frac{1+w}{1+w \cdot G_s} \cdot \rho_s \tag{3.1.13}$$

Equation (3.1.6) is called the volume–mass equation for unsaturated soil and is the most important equation which expresses the state of unsaturated soil by means of macroscopic physical quantities of void ratio, water content and degree of saturation. Figure 3.4(a) shows the twisted surface which prescribes the state of unsaturated soil in three-dimensional $e \sim w \sim S_r$ space. In other words, the state of unsaturated soil is prescribed by Equation (3.1.6) and can exist only in the twisted state surface shown in Figure 3.4(a). The twisted state surface is projected on $e \sim w$ plane, as shown in Figure 3.4(b). It is found from Figure 3.4(b) that the state of soil is limited to the colored trapezium area which is surrounded by e_{max} and e_{min} lines, and $e = G_s w$, where e_{max} is the maximum void ratio (= the minimum density) and e_{min} is the minimum void ratio (= the maximum density).

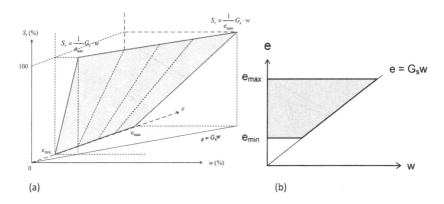

Figure 3.4 State surface of unsaturated soil (a) Twisted state surface in three dimensional e - w - S_r space, (b) State surface projected on e - w plane.

Figure 3.5(a) shows an example of shrinkage curve on the twisted state surface expressed by Equation (3.1.6). The shrinkage curve on the state surface is projected on the $e{\sim}w$ plane and then we can obtain a popular schematic diagram, as shown in Figure 3.5(b). Figure 3.6 shows another example of the compaction curve. Figure 3.6(a) shows plots on the twisted state surface. Figure 3.6(b) shows the plots projected on the $e{\sim}w$ plane in which the void ratio can be transformed into the dry density by using Equation (3.1.10), and then a popular compaction curve shown in Figure 3.6(c) is obtained.

It can be concluded from Equation (3.1.6) that the number of independent physical quantities which prescribe the state of unsaturated soil is two, arbitrarily selected from the void ratio, the water content and the degree of saturation, i.e., the combinations are (void ratio, water content), (void ratio, degree of saturation) and (water content, degree of saturation). It can also be concluded from Equation (3.1.12) that the number of independent physical quantities which prescribes the state of saturated soil is one, arbitrarily selected from the void ratio and the water content. In this book, the void ratio and the water content are selected as two fundamental and independent physical quantities for unsaturated coarse-grained soil. Therefore, it is shown that the other physical quantities defined by Equation (3.1.4) and Equations (3.1.7)–(3.1.11) can be derived from the void ratio and the water content. In geotechnical engineering practice, the wet density, the water content, and the densities of soil particles are directly measured by disturbed and/or undisturbed samples in the laboratory test. Then the void ratio is back calculated by substituting the density of soil particle, the water content and the wet density into the following equation derived from Equation (3.1.11):

$$e = \frac{(1+w)\rho_s}{\rho_t} - 1 \qquad\qquad (3.1.14)$$

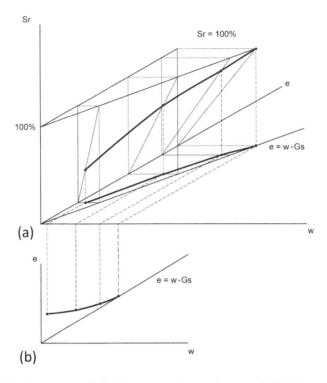

Figure 3.5 Shrinkage curve (a) Shrinkage curve in e-w-S_r space, (b) Shrinkage curve projected on e-w plane.

Figure 3.7 shows the compression curve of saturated soil and the soil water characteristic curve of unsaturated soil drawn in e-w-p space, where p denotes the pressure. The positive pressure is consolidation pressure and the negative pressure is suction. The compression curve can be drawn on the vertical plane with $e = wG_s$ in e-w-p space when the pressure changes in the positive range. The soil water characteristic curve can be drawn in e-w-s_u space when the pressure changes in the negative range. If the void ratio and the water content are adopted as the axes of fundamental and independent physical quantities and the third axis is adopted as the force or pressure, the mechanical behaviors of saturated and unsaturated soils might be synthetically shown in the same three-dimensional space without distinction, as shown in Figure 3.7.

3.2 MICROSCOPIC PROBABILISTIC MODELS OF SOLID, GAS, AND LIQUID PHASES

3.2.1 Elementary particulate body (EPB)

In Chapter 2, we have shown the triaxial specimen as a population, the "spoonful" cube as a sample population and the cube with a length of D_{cha}

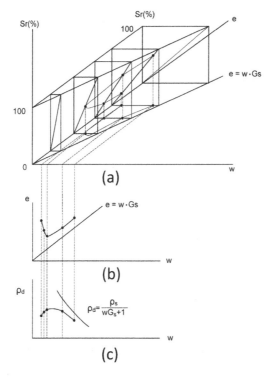

Figure 3.6 Compaction curve. (a) Plots in e-w-S_r space, (b) Projected plots on e-w plane, (c) Plots in e-ρ_d plane.

in Figure 2.3. Here a cubic sample which is taken from the sample population is called an elementary particulate body (EPB). D_{cha} is called the characteristic length and it depends on the grain size distribution and void ratio and is approximately D_{10} (diameter finer than 10%). The physical meaning of D_{cha} will be discussed in Section 4.2 later.

If D_{cha} is assumed to be 0.1 mm, which is common for coarse-grained soil, the number of EPB included in the triaxial specimen shown in Figure 2.3 is about 2×10^8, i.e., when the triaxial specimen and EPB are regarded as the population and sample respectively, it means that the population has about 2×10^8 samples. The following equation for the soil particle density can be derived from Equation (3.1.1), using Equation (2.4.1).

$$\rho_s = \frac{M_s}{V_s} = \frac{1}{N} \sum_{i=1}^{N} \frac{M_{s.i}}{V_{s.i}} = E\left[\frac{M_{s.i}}{V_{s.i}}\right] \tag{3.2.1}$$

where

N: number of samples (i.e., EPB),
$M_{s,i}$: mass of soil particles included in the i-th EPB,
$V_{s,i}$: volume of soil particles included in the i-th EPB.

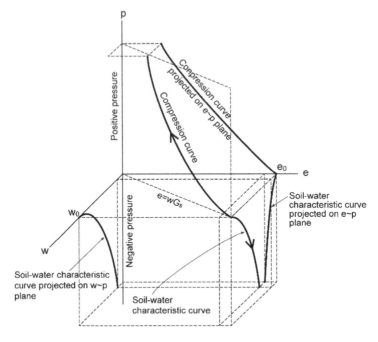

Figure 3.7 Schematic depiction of mechanical behaviors of saturated and unsaturated soil in 3-D space.

Similarly, the void ratio and the water content, which are the fundamental and independent physical quantities prescribed by the state of the soil, are obtained from Equations (3.1.2) and (3.1.3) respectively as follows:

$$e = \frac{V_v}{V_s} = \frac{1}{N} \sum_{i=1}^{N} \frac{V_{v.i}}{V_{s.i}} = E\left[\frac{V_{v,i}}{V_{s,i}}\right] \tag{3.2.2}$$

where $V_{v,i}$: volume of voids included in the i-th EPB.

$$w = \frac{M_w}{M_s} = \frac{1}{N} \sum_{i=1}^{N} \frac{M_{w,i}}{M_{s,i}} = E\left[\frac{M_{w,i}}{M_{s,i}}\right] = \frac{\rho_w}{\rho_s} E\left[\frac{V_{w,i}}{V_{s,i}}\right] \tag{3.2.3}$$

where $M_{w,i}$: mass of pore water included in the i-th EPB.

3.2.2 Modeling of an elementary particulate body (EPB)

In the grain size analysis of soil by means of both sieve analysis and sedimentation analysis, soil particles are assumed to be spherical in shape and the results are then applied to solve various problems in geotechnical engineering. In modeling of an EPB, a similar method is adopted, i.e., soil particles are assumed to be spherical in shape, although the shapes of soil particles are not spherical, but irregular. Furthermore, the soil particle is assumed to be rigid and uncrushable. As soil particles are rigid, the grain size distribution is not changed due to particle crush and/or abrasion at contact points of soil particles with the change in state of the soil block.

Figure 3.8 shows a contact point of two adjacent particles and a tangential plane formed at a contact point in the triaxial specimen shown Figure 2.3(a). The angles between the normal direction to the tangential plane and three axes, i.e., a set of $(\beta_1, \beta_2, \beta_3)$ are then defined as the contact angle at a contact point. The following equation is derived for the direction cosine:

$$\cos^2 \beta_1 + \cos^2 \beta_2 + \cos^2 \beta_3 = 1 \qquad\qquad (3.2.4)$$

Therefore, the number of independent contact angles is 2 of $(\beta_1, \beta_2, \beta_3)$, i.e., (β_1, β_2) or (β_2, β_3) or (β_1, β_3) are the contact angles in 3-D space as well as the volume-mass equation shown by Equation (3.1.6). In this book, consideration is limited to 2-D space and the contact angle is then defined as shown in Figure 3.8(b), i.e., the variable to identify the contact angle becomes β.

Figure 3.9(a) is the cube with the length D_{cha} shown in Figure 2.3(c). Two models are proposed to prescribe the state of the EPB, i.e., the models for particulate soil structure and pore structure as shown in Figure 3.9(b) and (c) respectively. For the particulate soil structure, the diameter of spherical soil particle D_s and the contact angle β, at the contact point of soil particles are adopted as independent random variables as shown in Figure 3.9(b).

Figure 3.9(c) shows the probabilistic model which can estimate the microscopic state of soil block and is called the elementary particulate model (EPM) for pore structure, where the pore in EPB is summed up to be

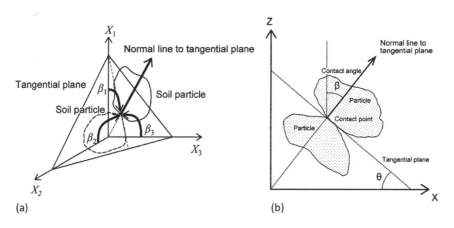

Figure 3.8 Definition of contact angle. (a) Contact angle defined in 3D space, (b) Contact angle defined in 2D space.

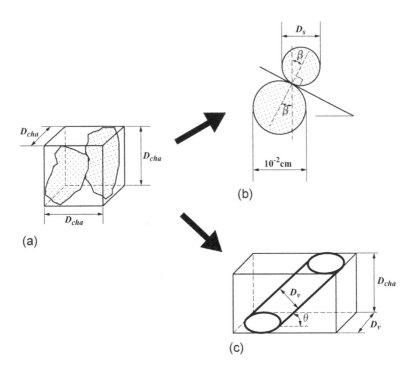

Figure 3.9 Modeling of soil particle structure and pore structure in elementary particulate body (EPB). (a) Elementary Particulate Body, (b) Modeling of Particulate Soil Structure, (c) Modeling of Pore Structure by Elementary Particulate Model (EPM).

the pipe with the diameter D_v and the inclination angle θ, which is considered to be the predominant flow direction of pore water and pore air. This modeling approach is similar to the modeling of the state of the soil block shown in Figure 3.2 by means of the phase diagram shown in Figure 3.3. Then D_v and θ are adopted as random variables for pore structure. The other parts, except those of the pipe, are regarded as the impermeable solid corresponding to the soil particles.

3.2.3 Modeling of particulate soil structure (solid phase)

It is well known that the grain size accumulation curve of most coarse non-organic soil can be approximately expressed by the cumulative grain size distribution function of logarithmic normal distribution. In this book, the following equation is used to express the grain size distribution of coarse-grained soil for the model of particulate soil structure shown in Figure 3.9(b), using Equations (2.3.6) and (2.5.10):

$$F_s(D_s) = \int_0^{D_s} f_{s(D_s)} \, dx = \int_0^{D_s} \frac{1}{\sqrt{2\pi}\zeta_s D_s} \exp\left(-\frac{(\ln D_s - \lambda_s)}{2\zeta_s^2}\right) dx \qquad (3.2.5)$$

where

D_s: diameter of soil particle assumed to be spherical,

$F_s(D_s)$:   cumulative grain size distribution function for coarse-grained soil,

$f_{s(Ds)}$:   grain size distribution function expressed by the logarithmic normal distribution, i.e.,

$$f_s(D_s) = \frac{1}{\sqrt{2\pi} \cdot \zeta_s \cdot D_s} \exp\left[-\frac{(\ln D_s - \lambda_s)^2}{2\zeta_s^2}\right] \qquad (3.2.6)$$

λ_s: mean value of $\ln D_s$ and

ζ_s: standard deviation of $\ln D_s$.

The values of distribution parameter, λ_s and ζ_s in Equation (3.2.6), are calculated by the regression analysis shown in Section 2.6.2 for the data obtained from the grain size analysis in the laboratory test. Then μ_s and σ_s can be calculated by using Equations (2.5.18) and (2.5.23). Note that λ_s corresponds to $\ln D_{50}$ and μ_s corresponds to the mean diameter of soil particles, i.e., the diameter finer than 50%, D_{50}, is never the mean diameter.

For another variable for the particulate soil structure shown in Figure 3.9(b), the probability density function of contact angle β, $f_\beta(\beta)$, is introduced, which is assumed to be a pentagonal shape, as shown in Figure 3.10, referring to the results of contact angle measured by the optical microscope examination of a sand specimen by Oda (1972). The pentagonal probability density function $f_\beta(\beta)$ is then expressed by the following equations:

$$\text{For } -\pi/2 \leq \beta \leq 0 \quad f_\beta(\beta) = \frac{2/\pi - 2 \cdot \varsigma_c}{\pi/2} \cdot \beta + \frac{2}{\pi} - \varsigma_c \qquad (3.2.7a)$$

$$\text{For } 0 \leq \beta \leq \pi/2 \quad f_\beta(\beta) = -\frac{2/\pi - 2 \cdot \varsigma_c}{\pi/2} \cdot \beta + \frac{2}{\pi} - \varsigma_c \qquad (3.2.7b)$$

where ς_c: distribution parameter which prescribes pentagonal shape.

Equation (3.2.7) means that the tangential plane at the contact point tends to be horizontal under a gravity field. In this book, the ratio of height at $\beta = \pm\pi/2$ to $\beta = 0$ is tentatively assumed to be 1:3 and then the distribution parameter ς_c is calculated to be the fixed value of 0.159, because it was found from the authors' past research work (Kitamura et al., 1998, Kitamura et al., 2012) that the distribution parameter ς_c in Equation (3.2.7) is not sensitive to the final results such as the soil water characteristic curve, the coefficient of water permeability, the safety factor, the

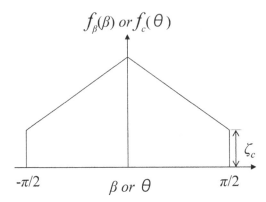

$f_\beta(\beta)$ or $f_c(\theta)$

$-\pi/2$ β or θ $\pi/2$

ζ_c

Figure 3.10 Probability density function of contact angle β and predominant flow direction θ.

stress–strain relation and so on, as shown in Chapters 6–10. This assumption will be verified by the development of the technology of measuring devices and data-processing, such as computed tomographic scanning (CT), image-processing owing to computer science, and so on in the near future. Tools such as the testing of hypotheses and the confidence intervals developed in the field of inferential statistics might also be promising means to verify the validity of the pentagonal probability density function with the distribution parameter ζ_c.

Consequently, the probabilistic state for the particulate soil structure can be estimated by Equation (3.2.6) for the particle diameter and Equation (3.2.7) for the contact angle.

3.2.4 Modeling of pore structure (gas and liquid phases)

3.2.4.1 Pore size distribution

There are several apparatuses used to examine the pore structure directly, such as the scanning electron microscope (SEM), the mercury intrusion porosimeter (MIP), and so on. However, in this subsection, a method to estimate the pore size distribution by means of grain size distribution function and the void ratio is proposed as the first approximate pore size distribution for the coarse-grained soil, because we aim to establish the versatile mechanical model for saturated–unsaturated coarse-grained soil based on only four fundamental soil properties, i.e., the grain size distribution function, the wet density of soil, the soil particle density and the pore water density. If the accurate pore structure can be easily obtained from simple experiments such as CT with image-processing in the future, the proposed approach to obtaining the pore size distribution in this subsection would be replaced by those direct approaches.

Figure 3.11 shows the elementary particulate model (EPM) with independent random variables D_v and θ, which has already been shown in Figure 3.9(c). Here we assume that the pore size distribution function for the coarse-grained soil is expressed by the logarithmic normal distribution as well as the grain size distribution function, though the bi-modal distribution is occasionally observed for the fine-grained soil. The following equation is obtained:

$$f_v(D_v) = \frac{1}{\sqrt{2\pi}\zeta_v \cdot D_v} \exp\left\{ -\frac{\left(\ln D_v - \lambda_v\right)^2}{2\zeta_v^2} \right\} \tag{3.2.8}$$

where

λ_v: mean value of $\ln D_v$ and

ζ_v: standard deviation of $\ln D_v$.

Furthermore, assuming that the coefficient of variation κ defined as Equation (2.4.14) for the pore size distribution is same as that for the grain size distribution function, the following equation is obtained:

$$\kappa = \frac{\sigma_s}{\mu_s} = \frac{\sigma_v}{\mu_v} \tag{3.2.9}$$

where

μ_s and σ_s: mean value and standard deviation of logarithmic normal distribution for the grain size distribution function and

μ_v and σ_v: mean value and standard deviation of logarithmic normal distribution for the pore size distribution.

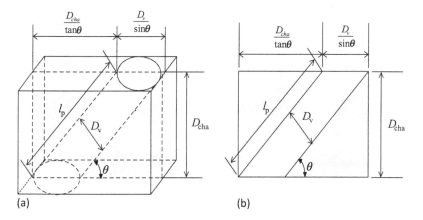

Figure 3.11 Elementary particulate model (EPM) (a) EPM in 3-Dimensional space, (b) EPM in 2-Dimensional space.

As described in Section 3.2.3, μ_s and σ_s are known from the soil test in the laboratory, and κ in Equation (3.2.9) is also known. Then ζ_v is known by using Equation (2.5.23). Consequently, an unknown parameter remaining in Equation (3.2.8) becomes only λ_v which is related to μ_v in Equation (3.2.9) by using Equation (2.5.18).

3.2.4.2 Distribution of predominant flow direction

Figure 3.12 shows schematically an element including two flat soil particles taken from the triaxial specimen, as shown in Figure 2.3. Let's consider the water or air flow in the direction of the x- or z- axis. When the water or air flow in the direction of the x-axis, it is easy to flow through the element. On the other hand, when the water or air flow in the z-axis, it is not easy to flow. From these properties of water or air flow through the element shown in Figure 3.12, we can deduce that the water or air easily flows in the direction parallel with the tangential plane formed at the contact point. Therefore, we can assume that the direction of predominant water or air flow is defined as the direction of tangential plane formed at the contact point. Furthermore, the idea of predominant flow direction can be applied to the EPB shown in Figure 3.9(a). In an EPB taken from the triaxial specimen as a sample, there is more than one contact point or there is no contact point. In the case of one contact point in an EPB, the predominant flow direction is parallel with the tangential plane. In the case of more than two contact points, the predominant flow direction is same as the average of direction of tangential planes. In the case of no contact point in an EPB, the flow through an EPB does not exist, i.e., this EPB is neglected. An EPB shown in Figure 3.9(a) can then be modeled by an EPM which is a cuboid with the height D_{cha} and consists of a pipe with the diameter D_v and the inclination angle θ, and other impermeable parts as shown in Figure 3.9(c). The pipe and other impermeable parts in an EPM correspond to the void and soil particles in an EPB respectively. As shown in Figure 3.8, the contact angle β, which is defined as the angle between the normal to

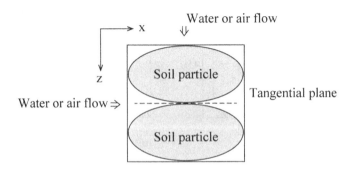

Figure 3.12 Predominant flow direction.

the tangential plane and the vertical line, is same as the angle between the tangential plane and the horizontal line. We then assume that the probability density function of inclination angle θ is same as the contact angle β as shown in Figure 3.8(b). The following equations are obtained for the probability density function of θ, which is the inclination angle of pipe in an EPM:

$$\text{For } -\pi/2 \leq \theta \leq 0 \quad f_c(\theta) = \frac{2/\pi - 2 \cdot \varsigma_c}{\pi/2} \cdot \theta + \frac{2}{\pi} - \varsigma_c \qquad (3.2.10(a))$$

$$\text{For } 0 \leq \theta \leq \pi/2 \quad f_c(\theta) = -\frac{2/\pi - 2 \cdot \varsigma_c}{\pi/2} \cdot \theta + \frac{2}{\pi} - \varsigma_c \qquad (3.2.10(b))$$

As shown in Figure 3.10, the probability density function of θ is same as that of β.

3.2.4.3 Estimation of λ_v from void ratio

The elementary particulate model (EPM) shown in Figure 3.11 is the microscopic model for pore structure of coarse-grained soil. Here the total and pipe volume of the EPM are denoted by V_{EPM} and $V_{\text{EPM},p}$ respectively, which correspond to $V_{s,i} + V_{v,i}$ and $V_{v,i}$ in Equation (3.2.2), i.e., these equations are obtained:

$$V_{\text{EPM}} = V_{s.i} + V_{v.i} \qquad (3.2.11)$$

$$V_{\text{EPM},p} = V_{v.i} \qquad (3.2.12)$$

Furthermore, V_{EPM} and $V_{\text{EPM},p}$ are obtained as follows, using the geometrical relations shown in Figure 3.11.

$$V_{\text{EPM}} = D_v \cdot \left(\frac{D_v}{\sin\theta} + \frac{D_{\text{cha}}}{\tan\theta} \right) \cdot D_{\text{cha}} \qquad (3.2.13)$$

$$V_{\text{EPM},p} = \pi \cdot \left(\frac{D_v}{2} \right)^2 \cdot \frac{D_{\text{cha}}}{\sin\theta} \qquad (3.2.14)$$

V_{EPM} in Equation (3.2.13) and $V_{\text{EPM},p}$ in Equation (3.2.14) can be expressed as the functions of random variables D_v and θ as follows:

$$V_{\text{EPM}} = \varphi_{\text{EPM}}(D_v, \theta) \qquad (3.2.15)$$

$$V_{\text{EPM},p} = \varphi_{\text{EPM},p}(D_v, \theta) \qquad (3.2.16)$$

where $\varphi_{\text{EPM}}(D_v, \theta)$ and $\varphi_{\text{EPM},p}(D_v, \theta)$: functions of D_v and θ for V_{EPM} and $V_{\text{EPM},p}$, corresponding to Equation (2.4.13).

The volume of the solid part in the EPM shown in Figure 3.11(b) is denoted by $V_{\text{EPM},s}$ and is obtained as follows, using Equations (3.2.11), (3.2.12), (3.2.15) and (3.2.16):

$$V_{\text{EPM},s} = \varphi_{\text{EPM}}(D_v,\theta) - \varphi_{\text{EPM},p}(D_v,\theta) = \varphi_{\text{EPM},s}(D_v,\theta) \tag{3.2.17}$$

where

$V_{\text{EPM},s}$:   volume of solid part in EPM and
$\varphi_{\text{EPM},s}(D_v,\theta)$: functions of D_v and θ for $V_{\text{EPM},s}$.

Using Equations (2.4.12), (3.2.2), (3.2.16) and (3.2.17), the following equation can be obtained:

$$e = \int_0^{\infty} \int_{-\frac{\pi}{2}}^{\frac{\pi}{2}} \frac{\varphi_{EPM,p(D_v,\theta)}}{\varphi_{EPM,s}(D_v,\theta)} \cdot f_v(D_v) \cdot f_c(\theta) d\theta dD_v \tag{3.2.18}$$

As the void ratio on the left side of Equation (3.2.18), the distribution parameter ζ_v of $f_v(D_v)$ and the distribution parameter ζ_c of $f_c(\theta)$ are known, the unknown parameter becomes only λ_v of $f_v(D_v)$. Therefore, the distribution parameter λ_v is numerically back calculated to satisfy Equation (3.2.18). Consequently, it is found that the distribution parameter λ_v depends on the void ratio.

Finally, note that the maximum void ratio calculated by the elementary particulate model (EPM) is 3.66 (=$\pi/(4-\pi)$) for the case of $\theta = \pm\pi2$. It means that the void ratio of the sample (EPB) taken out of the triaxial specimen is assumed to be less than 3.66, although the samples with more than 3.66 might really be included in the soil block (sample population). The validity of this assumption should be investigated by applying the observation and image-processing technologies with respect to microscopes and other devices in the field of computer science, electrical engineering, etc. in the near future.

3.2.4.4 Estimation of threshold value d_w from water content

In the drying process of saturated soil, the entry of air into pores filled with water firstly begins in larger pores, and inversely, in the wetting process of unsaturated soil, the entry of water into pores filled with air begins in smaller pores. Applying these behaviors to the elementary particulate model (EPM), it can be assumed that the pipe with the range of diameter $0 < D_v \leq d_w$ is filled with water and the pipe with the range of diameter $d_w < D_v < \infty$ is filled with air for the unsaturated soil. The following equation can then be derived by using Equation (2.4.13), (3.2.3), (3.2.16) and (3.2.17) in a similar way to the derivation of Equation (3.2.18):

$$w = \frac{\rho_w}{\rho_s} \cdot \int_0^{d_w} \int_{-\frac{\pi}{2}}^{\frac{\pi}{2}} \frac{\varphi_{EPM,p(D_v,\theta)}}{\varphi_{EPM,s}(D_v,\theta)} \cdot f_v(D_v) \cdot f_c(\theta) d\theta dD_v \qquad (3.2.19)$$

Equation (3.2.19) is rewritten as follows:

$$w \cdot G_s = \int_0^{d_w} \int_{-\frac{\pi}{2}}^{\frac{\pi}{2}} \frac{\varphi_{EPM,p}(D_v,\theta)}{\varphi_{EPM,s}(D_v,\theta)} \cdot f_v(D_v) \cdot f_c(\theta) d\theta dD_v \qquad (3.2.20)$$

It is found that Equation (3.2.20) is the microscopic description of the volume–mass equation, Equation (3.1.6), where the microscopic physical quantity d_w corresponds to the macroscopic physical quantity S_r, i.e., when $S_r = 1$, $d_w = \infty$.

As the water content in the left side of Equation (3.2.19), the distribution parameters λ_v and ζ_v of $f_v(D_v)$ and the distribution parameter ζ_c of $f_c(\theta)$ are known, the unknown parameter remained becomes only d_w in Equation (3.2.19). Therefore d_w is numerically back calculated to satisfy Equation (3.2.19). It is found that the diameter d_w is the threshold of the pipe in the EPM shown in Figure 3.11 where the pipe with the diameter larger than d_w is occupied with air and the pipe smaller than d_w is occupied with water.

Figure 3.13 shows the relations between the macroscopic physical quantities (grain size distribution, void ratio and water content) and the microscopic physical quantities (characteristic length, pore size distribution and threshold value d_w). The microscopic physical quantities are numerically obtained from the macroscopic physical quantities which are obtained from

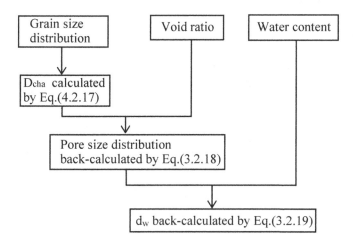

Figure 3.13 Numerical calculation procedure to relate macroscopic physical quantities to microscopic ones.

the popular soil tests in the laboratory. Firstly, the characteristic length D_{cha} of EPB shown in Figure 3.9(a) is calculated by using the grain size distribution function which will be discussed in Section 4.2 later. Secondly, substituting the measured void ratio and grain size distribution function to Equation (3.2.18), the parameter λ_v of pore size distribution in Equation (3.2.8) is back-calculated. Then, substituting the measured water content to Equation (3.2.19), the threshold value d_w which is the maximum diameter of pipe occupied with pore water in EPM shown in Figure 3.11 is obtained.

3.2.4.5 Summary

The microscopic modeling of particulate soil structure and pore structure for coarse-grained soil discussed in this section is concluded as follows:

- The microscopic state of particulate soil structure in soil can be estimated by Equations (3.2.6) and (3.2.7), and the microscopic state of pore structure in soil can be estimated by Equations (3.2.8) and (3.2.10).
- All of the parameters included in this section except $\zeta_c = 0.159$ in Equations (3.2.8) and (3.2.10) can be determined by the densities of soil particle and water, the grain size distribution function, the void ratio and the water content which are obtained from soil tests in the laboratory.
- When the void ratio of EPB which is a sample taken out of triaxial specimen is more than 3.66, this sample is neglected.

REFERENCES

Haruyama, M. (1969). Effect of water content on the shear characteristics of granular soils such as Shirasu. *Soils and Foundations*, 9(3), 35–57.
Kitamura, R., Fukuhara, S., Uemura, K., Kisanuki, J. and Seyama, M. (1998). A numerical model for seepage through unsaturated soil. *Soils and Foundations*, 38(4), 261–265.
Kitamura, R., Yamada, M., Kawabata, K., Inagaki, Y. and Araki, K. (2012). Mechanical & numerical model for deformation behaviour of unsaturated soil. *Journal of JSCE*, 68(2), I_487–492 (in Japanese).
Oda, M. (1972). Initial fabrics and their relations to mechanical properties of granular material. *Soils and Foundations*, 12(1), 17–36.

Chapter 4

Microscopic physical quantities derived from void ratio and probability distributions

4.1 NUMBER OF SOIL PARTICLES PER UNIT VOLUME

Let's consider a unit cube of a soil block with the void ratio e as shown in Figure 2.3(b). In the phase diagram, the unit cube with the void ratio e is modeled as shown in Figure 4.1. When the mass of soil particles in the unit cube of soil block is denoted by $M_{s,\text{unit}}$, the following equation is obtained, referring to Figure 4.1:

$$M_{s,\text{unit}} = \frac{1}{1+e}\rho_s \tag{4.1.1}$$

where

$M_{s,\text{unit}}$:   mass of soil particles in the unit cube of the soil block, which is same as the dry density shown in Equation (3.1.10),

e: void ratio of unit cube and

ρ_s: density of soil particles.

Figure 4.2 shows a cumulative grain size distribution function of the soil block expressed by Equation (3.2.5). The vertical axis (percent finer by mass) or the horizontal axis (normal scale) is divided into n equal intervals (e.g., $n = 360$). In this section, the explanation is shown for the case where the vertical axis is divided as shown in Figure 4.2.

The diameter and the interval corresponding to the i-th interval of vertical axis are respectively denoted by $D_{s,i}$ and $\Delta D_{s,i}$. The mass of soil in the i-th interval is denoted by $M_{s,\text{unit},i}$, and then the following equation is obtained:

$$M_{s,\text{unit}} = \sum_{i=1}^{n} M_{s,\text{unit},i} \tag{4.1.2}$$

where

$M_{s,\text{unit},i}$:   mass of soil particles in the i-th interval and

n: division number of cumulative grain size distribution function.

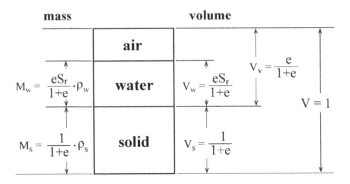

Figure 4.1 Phase diagram in a case where a whole soil block is the unit volume.

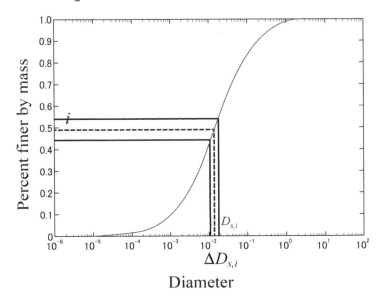

Figure 4.2 Cumulative grain size distribution divided into n parts of vertical axis and the i-th division with corresponding diameter $D_{s,i}$.

When the number of soil particles included in the i-th interval is denoted by $N_{prt,i}$, the following equation is obtained:

$$M_{s,unit,i} = N_{prt,i} \cdot \frac{D_{s,i}^3}{6} \pi \cdot \rho_s \qquad (4.1.3)$$

where $N_{prt,i}$: number of soil particles included in the i-th interval.

Applying Equations (2.4.8) and (3.2.5) to relate $M_{s,unit,i}$ to $M_{s,unit}$, the following equation is obtained:

$$M_{s,unit,i} = M_{s,unit} \cdot f_s(D_{s,i}) \cdot \Delta D_{s,i} \qquad (4.1.4)$$

where $f_s(D_{s,i})$: probability density function obtained from the grain size accumulation curve (see Equation (3.2.6).

Substituting Equations (4.1.1) and (4.1.3) into Equation (4.1.4), the following equation is obtained:

$$N_{prt,i} \cdot \rho_s \frac{D_{s,i}^3}{6} \pi = \frac{1}{1+e} \rho_s \cdot f_s(D_{s,i}) \cdot \Delta D_{s,i} \tag{4.1.5}$$

Equation (4.1.5) is rewritten as follows:

$$N_{prt,i} = \frac{1}{1+e} \cdot \frac{6}{D_{s,i}^3 \cdot \pi} \cdot f_s(D_{s,i}) \cdot \Delta D_{s,i} \tag{4.1.6}$$

As the void ratio, the grain size distribution and the density of soil particles in the right side of Equation (4.1.6) are known, $N_{prt,i}$ is determined for a given $D_{s,i}$.

The number of soil particles per unit volume N_{prt} is related to $N_{prt,i}$ in the i-th interval as follows:

$$N_{prt} = \sum_{i=1}^{n} N_{prt,i} \tag{4.1.7}$$

Substituting Equation (4.1.6) into Equation (4.1.7), the following equation is obtained:

$$N_{prt} = \frac{1}{1+e} \cdot \frac{6}{\pi} \sum_{i=1}^{n} \frac{1}{D_{s,i}^3} \cdot f_s(D_{s,i}) \cdot \Delta D_{s,i} = \frac{1}{1+e} \cdot \frac{6}{\pi} \cdot \alpha_{grain} \tag{4.1.8}$$

where

$$\alpha_{grain} = \sum_{i=1}^{n} \frac{1}{D_{s,i}^3} \cdot f_s(D_{s,i}) \cdot \Delta D_{s,i} \tag{4.1.9}$$

It is found from Equation (4.1.8) that the number of soil particles per unit volume, N_{prt}, can be numerically calculated by using the grain size distribution and the void ratio of the soil block.

4.2 CHARACTERISTIC LENGTH

4.2.1 Brief review of microscopic interpretation of effective stress used in the conventional saturated soil mechanics

Figure 4.3 shows a conceptual state to commonly explain the microscopic physical meaning of effective stress of saturated soil in the textbook of conventional soil mechanics, although it is much more difficult for this state to

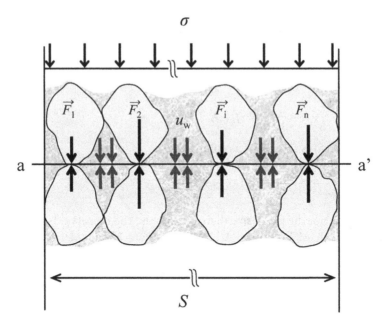

Figure 4.3 Conceptual state of soil to explain effective stress microscopically.

exist in real soil. Considering the force equilibrium on the hypothetical a–a' section, the following equation is derived.

$$\sigma \cdot S = \sum_{i=1}^{n} F_i + u_w \left(S - \sum_{i=1}^{n} a_i \right)$$

(4.2.1)

where

σ: stress generated on the a–a' plane,
S: area of cross section on the a–a' plane,
F_i: inter-particle force at the i-th contact point on the a–a' plane,
u_w: pore water pressure,
a_i: contact area of the i-th contact point,
n: number of contact points on the a–a' plane.

Note that the self-weight of soil particles is neglected in Equation (4.2.1).

Dividing both side of Equation (4.2.1) by S, the following equation is obtained:

$$\sigma = \sum_{i=1}^{n} \left(\frac{F_i}{S} \right) + u_w \left(1 - \sum_{i=1}^{n} \left(\frac{a_i}{S} \right) \right)$$

(4.2.2)

Assuming $a_i \cong 0$, Equation (4.2.2) is rewritten as follows:

$$\sigma = \sum_{i=1}^{n} \left(\frac{F_i}{S} \right) + u_w \tag{4.2.3}$$

The popular effective stress equation of saturated soil is then obtained as follows:

$$\sigma = \sigma' + u_w \tag{4.2.4}$$

where

$$\sigma' = \sum_{i=1}^{n} \left(\frac{F_i}{S} \right) \tag{4.2.5}$$

Using Equation (2.4.2), Equation (4.2.5) is rewritten as follows:

$$\sigma' = \sum_{i=1}^{n} \left(\frac{F_i}{s} \right) = \frac{n}{S} \cdot E[F_i] \tag{4.2.6}$$

where $E[F_i]$: mean value of $F_i (i = 1 \sim n)$.

It is found from Equation (4.2.6) that the conventional effective stress for saturated soil is the mean value of inter-particle force multiplied by the number of contact points per unit area.

4.2.2 Derivation of characteristic length

Figure 4.4(a) shows the simple cubic packing of 16 uniform spheres, i.e., the numbers of column ($=i$) and row ($=j$) are 4 respectively. The density and diameter of particles are denoted by ρ_s and D_s respectively.

Here let's consider the force due to gravitational force per unit area acted on the bottom plane. As there are four spheres in a column, i.e., $j = 4$, the force acted on the bottom plane of the i-th column, F_i, is obtained as follows:

$$F_i = \sum_{j=1}^{4} \rho_s g \cdot \frac{D_{s,j}^3 \cdot \pi}{6} = 4\rho_s g \cdot \frac{D_s^3 \cdot \pi}{6} \tag{4.2.7}$$

The average force acted on the bottom plane is obtained as follows, using Equation (2.4.1):

$$E[F_i] = \frac{1}{4} \sum_{i=1}^{4} F_i \tag{4.2.8}$$

where $E[F_i]$: average force of the i-th column ($i = 1 \sim 4$) acted on the bottom plane,

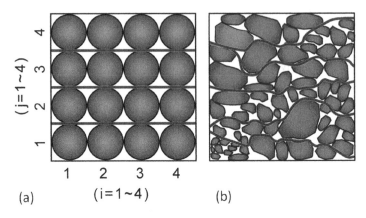

(a)   (i=1~4)   (b)

Figure 4.4 Contact planes for uniform particles and random particles in shape and size (a) Simple cubic packing of uniform spheres, (b) Random packing of irregular particles.

Substituting Equation (4.2.7) into Equation (4.2.8), the following equation is obtained:

$$E[F_i] = \frac{1}{4}\sum_{i=1}^{4} 4\rho_s g \cdot \frac{D_s^3 \cdot \pi}{6} = 4\rho_s g \cdot \frac{D_s^3 \cdot \pi}{6} \tag{4.2.9}$$

The resultant force per unit area (= pressure) at the bottom plane p_{bottom} is then obtained as follows, in a similar way to effective stress described in Section 4.2.1.

$$p_{\text{bottom}} = \frac{\left(\text{Weight of spheres in a single row}\right)\times\left(\text{Number of rows}\right)}{\text{Area}}$$

$$= \frac{\left(4\rho_s g \cdot \frac{D_s^3}{6}\right)\times(4)}{\left(4D_s \cdot D_s\right)} = \rho_s g \cdot \frac{2}{3}\pi D_s \tag{4.2.10}$$

where p_{bottom}: pressure acted on the bottom plane.

The plane formed by the contact points is flat for the simple cubic packing as shown in Figure 4.4(a). On the other hand, the plane connected by the contact points of a real soil block is not flat but curved, as shown in Figure 4.4(b), because of the irregularity in the shape and size of soil particles and the randomness of their arrangement. To obtain the resultant force per unit area of the bottom plane for the soil block, as well as the simple cubic packing, it is necessary for the soil block to determine the number of rows in Equation (4.2.10). Therefore, we must introduce a bold new parameter called the characteristic length D_{cha} for soil blocks with irregularly shaped particles. Here we hypothesize that the characteristic length D_{cha} can be defined as the diameter of uniform sphere particles of simple cubic packing to determine the number of rows for the soil block shown in Figure 4.4(b). Furthermore, the number of particles per unit volume with the diameters larger than D_{10} (=diameter finer than 10% in the grain size accumulation curve) denoted by $N_{prt>10}$ is used to calculate the number of rows. D_{10}, called the effective grain size, might be one of the key parameters to govern the mechanical behavior of coarse-grain soil because D_{10} is used in the equation for the coefficient of uniformity for the grain size accumulation curve and in the empirical Hazen's equation for the permeability coefficient. Therefore, it might be reasonable to introduce the microscopic physical quantity $N_{prt>10}$ to determine D_{cha}. The following equation is then derived in a manner similar to Equation (4.2.10):

$$p_{bottom} = \rho_s \cdot g \cdot \frac{D_{cha}^3 \pi}{6} \cdot N_{prt>10} \qquad (4.2.11)$$

Referring to Equations (4.1.8) and (4.1.9), the following equation is obtained for $N_{prt>10}$:

$$N_{prt>10} = \sum_{i=n_{>10}}^{n} N_{prt,i} = \frac{1}{1+e} \cdot \frac{6}{\pi} \cdot \sum_{i=n_{>10}}^{n} \frac{1}{D_{s,i}^3} \cdot f_s(D_{s,i}) \cdot \Delta D_{s,i}$$

$$= \frac{1}{1+e} \cdot \frac{6}{\pi} \cdot \alpha_{grain>10} \qquad (4.2.12)$$

where

$n_{>10}$: number of intervals in the range larger than D_{10}.

$$\alpha_{grain>10} = \sum_{i=n_{>10}}^{n} \frac{1}{D_{s,i}^3} \cdot f_s(D_{s,i}) \cdot \Delta D_{s,i} \qquad (4.2.13)$$

$f_s(D_{s,i})$: probability density function obtained from grain size accumulation curve (see Equation (3.2.6)).

On the other hand, the average force per unit area at the bottom plane of the unit cube (i.e., the pressure due to gravitational force) is obtained by using the dry density and Equation (3.1.8) as follows:

$$p_{bottom} = \rho_d \cdot g \cdot 1 = \frac{1}{1+e} \rho_s \cdot g \qquad (4.2.14)$$

where ρ_d: dry density of soil.

The following equation is obtained from Equations (4.2.12) and (4.2.14):

$$\frac{1}{1+e} \rho_s \cdot g = \rho_s \cdot g \cdot \frac{D_{cha}^3 \pi}{6} \cdot N_{prt>10} \qquad (4.2.15)$$

Substituting Equation (4.2.12) into Equation (4.2.15), the following equation is obtained:

$$\frac{1}{1+e} \cdot \rho_s \cdot g = \rho_s \cdot g \cdot \frac{D_{cha}^3 \pi}{6} \cdot \frac{1}{1+e} \cdot \frac{6}{\pi} \cdot \alpha_{grain>10} \qquad (4.2.16)$$

Then D_{cha} is obtained from Equation (4.2.16) as follows:

$$D_{cha} = \left(\frac{1}{\alpha_{grain>10}} \right)^{\frac{1}{3}} \qquad (4.2.17)$$

Figure 4.5(a) shows an image for the microscopic state of real soil with contact points. We can draw lines connecting contact points to make paths similar to the a–a' line shown in Figure 4.3. Figure 4.5(b) shows curved paths connected through contact points of particles with irregular size and shape. Figure 4.5(c) shows the hypothesized flat surfaces and the spheres with the diameter of characteristic length D_{cha} between them.

It is found from Equation (4.2.17) that the characteristic length D_{cha} is one of the material constants for a given soil, derived from the grain size accumulation curve as well as the uniformity coefficient and the coefficient of curvature used in the conventional soil mechanics.

It may be interpreted from Equation (4.2.15), and Figures 4.5(c) that the characteristic length D_{cha} is the parameter to average the random thickness of row as shown in Figure 4.5(a) and (b), and to link the body force with the surface force for the particulate material. Furthermore, it may be considered that the characteristic length D_{cha} is the diameter of uniform spheres with simple cubic packing which is derived from the grain size distribution of the real soil as shown in Section 4.4.

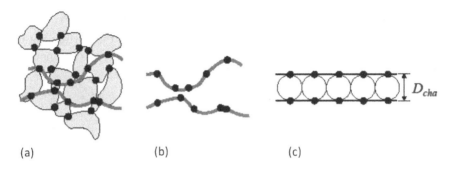

(a)   (b)   (c)

Figure 4.5 Physical meaning of D_{cha} (a) Curved path connected with contact points, (b) Thickness of layer through path, (c) Flat plane corresponding to curved path.

4.3 NUMBERS OF CONTACT POINTS PER UNIT VOLUME AND UNIT AREA

Let's imaging the cube with unit volume and unit area as shown in Figure 2.3(b). It is then assumed that any unit cube of sample population taken out from the triaxial specimen has the same particulate soil structure and pore structure, i.e., the triaxial specimen is uniform, although the specimen is not uniform really. Furthermore, it is assumed that the number of contact points per unit area does not depend on the position and angle of the plane in the unit cube. To prove these assumptions, the testing of the statistical hypothesis is theoretically needed in the inferential statistics, but this procedure is not carried out in this book.

Figure 4.6 shows a schematic state of soil particles where six contact points are formed at the central particle with its surrounding particles. Field (1963) proposed the following empirical equation to estimate the number of contact points per particle for the coarse-grained soil:

$$N_{crd} = \frac{12}{1+e} \tag{4.3.1}$$

where

N_{crd}: number of contact points per particle (co-ordination number) and
e: void ratio.

Oda (1977) experimentally obtained the relation between the number of contact points per particle and the void ratio for a homogeneous assembly and two-mixed assembly, as shown in Figure 4.7 where

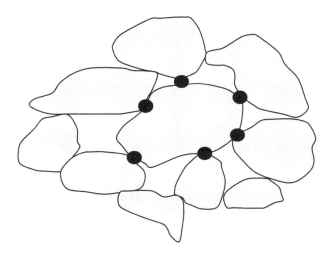

Figure 4.6 Contact points per particle.

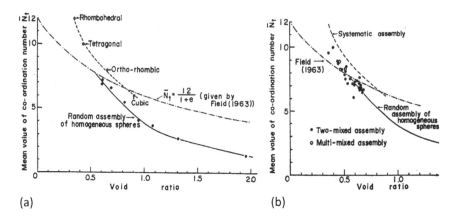

Figure 4.7 Relation between the number of contact points per particle and the void ratio (after Oda, 1977) (a) Relation between the mean value of co-ordination number and the void ratio in the homogeneous assemblies, (b) Relation between the mean value of co-ordination number and the void ratio in the two-mixed and multi-mixed assemblies.

the relations of Equation (4.3.1) are included. Based on these previous experimental results, we tentatively use Equation (4.3.1) to calculate the number of contact points per unit volume, although the number of contact points per particle depends on not only the void ratio, but also the shape and size of the soil particles. When measuring devices and data-processing technology for the number of contact points per particle – such as computed tomographic scanning (CT), imageprocessing and so on – are available in the future, owing to the development of computer science, Equation (4.3.1) will be replaced by a more accurate one.

Using Equation (4.3.1), the number of contact points per unit volume, N_{cv} can be calculated by the following equation:

$$N_{cv} = \frac{N_{crd}}{2} \cdot N_{prt} \tag{4.3.2}$$

where

N_{cv}: number of contact points per unit volume and
N_{prt}: number of soil particles per unit volume.

Substituting Equation (4.1.8) into Equation (4.3.2), the following equation is obtained:

$$N_{cv} = \frac{36}{(1+e)^2 \cdot \pi} \cdot \left(\sum_{i=1}^{n} \frac{1}{D_{s,i}^3} \cdot f_s(D_{s,i}) \cdot \Delta D_{s,i} \right) = \frac{36}{(1+e)^2 \cdot \pi} \cdot \alpha_{grain} \tag{4.3.3}$$

Here let's consider the unit cube, which is divided into layers with the thickness of D_{cha}, as shown in Figure 4.8. The number of layers for the unit cube is obtained as follows:

$$N_{layer} = \frac{1}{D_{cha}}$$
(4.3.4)

where
N_{layer}: number of layers and
D_{cha}: characteristic length expressed by Equation (4.2.17).

We then assume that the number of contact points per unit area can be related to the number of contact points per unit volume by the following equation:

$$N_{ca} = \frac{N_{cv}}{N_{layer}}$$
(4.3.5)

where N_{ca}: number of contact points per unit area.

Substituting Equation (4.3.4) into Equation (4.3.5), the following equation is obtained:

$$N_{ca} = N_{cv} \cdot D_{cha}$$
(4.3.6)

And substituting Equations (4.2.17) and (4.3.3), the following equation is obtained:

$$N_{ca} = \frac{36}{(1+e)^2 \cdot \pi} \cdot \frac{\alpha_{grain}}{\left(\alpha_{grain>10}\right)^{1/3}}$$
(4.3.7)

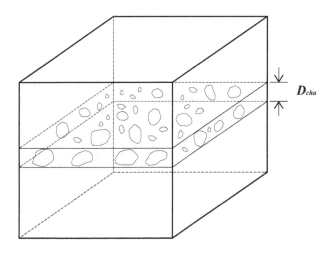

Figure 4.8 Layers with thickness D_{cha} in a cube with unit length.

It is concluded in this section that the contact points per unit volume and unit area can be calculated by Equations (4.3.3) and (4.3.7), i.e., the void ratio and the grain size distribution are only necessary to calculate the numbers of contact points per unit volume and unit area.

4.4 CALCULATION OF N_{PRT}, D_{CHA}, N_{CV}, AND N_{CA} FOR SIMPLE CUBIC PACKING OF UNIFORM SPHERES

In this section, the physical quantities derived from the geometrical relation for the simple cubic packing of uniform spheres is compared with those calculated by the equations derived from the previous sections in Chapter 4.

4.4.1 N_{prt} and void ratio derived from geometrical relation of simple cubic packing

Let's consider a simple cubic packing of uniform spheres in a cubic container, as shown in Figure 4.9, to check the validity of equations derived in the previous sections, where 64 (=4×4×4) uniform spheres with diameter D_s are included in a cubic container. The volume of cubic container V and total number of spheres N_s are as follows:

$$V = \left(4D_s\right)^3 = 64D_s^3 \tag{4.4.1}$$

$$N_s = 64 \tag{4.4.2}$$

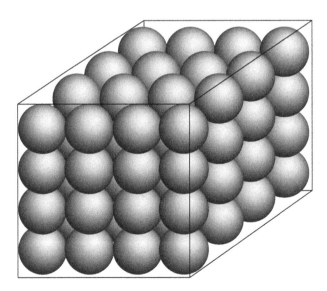

Figure 4.9 Simple cubic packing of uniform spheres.

Then the number of spheres per unit volume is obtained as follows, using Equations (4.4.1) and (4.4.2):

$$N_{prt} = \frac{N_s}{V} = \frac{64}{64D_s^3} = \frac{1}{D_s^3} \tag{4.4.3}$$

The volume of spheres is as follows:

$$V_s = 64 \times \frac{\pi}{6} D_s^3 \tag{4.4.4}$$

Substituting Equations (4.4.1) and (4.4.4) into Equation (3.1.2), the following equation is obtained:

$$e = \frac{V_v}{V_s} = \frac{V - V_s}{V_s} = \frac{6 - \pi}{\pi} \tag{4.4.5}$$

The following equation is then obtained, using Equation (4.4.5):

$$1 + e = \frac{6}{\pi} \tag{4.4.6}$$

4.4.2 Calculation of N_{prt}

α_{grain} defined by Equation (4.1.9) and $\alpha_{grain>10}$ defined by Equation (4.2.13) are rewritten as follows for the simple cubic packing of uniform spheres:

$$\alpha_{grain} = \alpha_{grain>10} = \frac{1}{D_s^3} \tag{4.4.7}$$

Substituting Equations (4.4.6) and (4.4.7) into Equation (4.1.8), the following equation is obtained:

$$N_{prt} = \alpha_{grain} = \frac{1}{D_s^3} \tag{4.4.8}$$

Comparing Equation (4.4.8) with Equation (4.4.3), it is found that Equation (4.1.8) is also applicable to the simple cubic packing of uniform spheres.

4.4.3 Calculation of D_{cha}

Substituting Equation (4.4.7) into Equation (4.2.17), the following equation is obtained:

$$D_{cha} = \left(\frac{1}{\alpha_{grain>10}} \right)^{\frac{1}{3}} = \left(\frac{1}{\frac{1}{D_s^3}} \right)^{\frac{1}{3}} = D_s \tag{4.4.9}$$

It is found from Equation (4.4.9) that the characteristic length for simple cubic packing of uniform spheres becomes the diameter of a sphere.

4.4.4 Calculation of N_{cv} and N_{ca}

Substituting Equations (4.4.6) and (4.4.7) into Equation (4.3.3), the following equation is obtained:

$$N_{cv} = \pi \times \frac{1}{D_s^3} = \pi \times \frac{1}{D_s} \times \frac{1}{D_s} \times \frac{1}{D_s} \tag{4.4.10}$$

$1/D_s$ in Equation (4.4.10) means the number of spheres in the unit length of the cube.

Substituting Equations (4.4.6) and (4.4.7) into Equation (4.3.7), the following equation is obtained:

$$N_{ca} = \pi \cdot \frac{\dfrac{1}{D_s^3}}{\dfrac{1}{D_s}} = \pi \times \frac{1}{D_s} \times \frac{1}{D_s} \tag{4.4.11}$$

The number of contact points per particle is exactly 6 for the simple cubic packing of uniform spheres, i.e., $N_{crd} = 6$, but the following value is obtained when we use Equation (4.3.1):

$$N_{crd} = 2\pi = 6.3 \tag{4.4.12}$$

Comparing 6.3 in Equation (4.4.12) with the exact number of 6 for the simple cubic packing of uniform spheres, it is deduced that the small errors are hidden in the number of contact points per a real soil particle estimated by Equation (4.3.1), but it may be acceptable to use Equation (4.3.1) to estimate the approximate number of contact points per particle.

REFERENCES

Field, W. G. (1963). Towards the statistical definition of a granular mass. *Proc. 4th A and N.Z. Conf. on SMFE*, 143–148.
Oda, M. (1977). Co-ordination number and its relation to shear strength of granular material. *Soils and Foundations*, 17(2), 29–42.

Chapter 5

Inter-particle force vectors and inter-particle stress vectors

In this chapter, we will discuss the inter-particle force vector generated at a contact point and the inter-particle stress vector on a plane which is defined as an average of inter-particle force vectors per unit area, relating to the particulate soil structure shown in Figure 3.9(b).

5.1 NOTATION OF INTER-PARTICLE FORCE VECTOR AND INTER-PARTICLE STRESS VECTOR

The notation of inter-particle force vector and inter-particle stress vector is shown in this section.

The inter-particle force is generated at a contact point formed by adjacent two particles, as shown in Figure 5.1. The inter-particle force vector at the i-th contact point of tangential plane with the inclination angle β is denoted by $\vec{F}_{\beta,i}$, which is expressed by the unit vectors and vector components as follows:

$$\vec{F}_{\beta,i} = F_{\beta,i,N} \cdot \vec{e}_{\beta,i,N} + F_{\beta,i,T} \cdot \vec{e}_{\beta,i,T} \qquad (5.1.1)$$

where

$\vec{F}_{\beta,i}$: inter-particle force vector with the contact angle β at the i-th contact point,

$\vec{e}_{\beta,i,N}$: unit vector in the normal direction to the plane with inclination angle β at the i-th contact point,

$\vec{e}_{\beta,i,T}$: unit vector in the tangential direction to the plane with inclination angle β at the i-th contact point,

$F_{\beta,i,N}$: normal component of inter-particle force vector $\vec{F}_{\beta,i}$, which is scalar quantity and

$F_{\beta,i,T}$: tangential component of inter-particle force vector $\vec{F}_{\beta,i}$, which is scalar quantity.

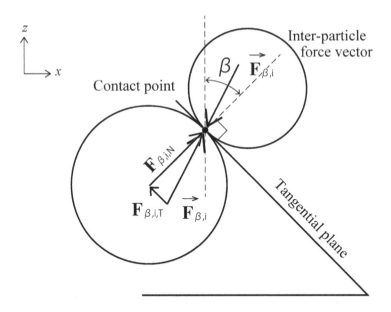

Figure 5.1 Two particles in Figure 3.9(b), taken from triaxial soil specimen in Figure 2.3.

The inter-particle stress vector generated on the plane with the inclination angle β is denoted by $\vec{F}_\beta$ which is expressed by the unit vectors and vector components as well as Equation (5.1.1) as follows:

$$\vec{F}_\beta = F_{\beta,N} \cdot \vec{e}_{\beta,N} + F_{\beta,T} \cdot \vec{e}_{\beta,T} \tag{5.1.2}$$

where

$\vec{F}_\beta$: inter-particle stress vector on the plane with the inclination angle β,

$\vec{e}_{\beta,N}$: unit vector in the normal direction to the plane with inclination angle β,

$\vec{e}_{\beta,T}$: unit vector in the tangential direction to the plane with inclination angle β,

$F_{\beta,N}$: normal component of inter-particle stress vector $\vec{F}_\beta$, which is scalar quantity and

$F_{\beta,T}$: tangential component of inter-particle stress vector $\vec{F}_\beta$, which is scalar quantity.

5.2 INTER-PARTICLE FORCE VECTOR AT A CONTACT POINT AND INTER-PARTICLE STRESS VECTOR ON A PLANE

5.2.1 Inter-particle force vector

Figure 5.1 shows two spherical particles that are considered to be a model of real soil particles taken from the triaxial soil specimen (population) or the cube on the spoon (sample population) shown in Figures 2.3(a) and (b) respectively. In other words, the two spherical particles shown in Figure 5.1 are regarded as a model of particulate soil structure shown in Figure 3.9(b).

The inter-particle force vector $\vec{F}_{\beta,i}$ shown in Figure 5.1 is defined as the vector generated at the i-th contact point of two adjacent particles with the contact angle β under a given pressure condition of triaxial testing apparatus. The inter-particle force vectors at all contact points in the soil block have wide-ranging amounts and various directions. These inter-particle force vectors in the soil block are generated by the self-weight of particle, capillary force, seepage force, external force, osmotic pressure and physicochemical action. Therefore the inter-particle force vectors generated at the i-th contact point with the contact angle β can be generally expressed as follows:

$$\vec{F}_{\beta,i} = \vec{F}_{\beta,\text{grav},i} + \vec{F}_{\beta,\text{seep},i} + \vec{F}_{\beta,\text{matr},i} + \vec{F}_{\beta,\text{ext},i} + \vec{F}_{\beta,\text{osmo},i} + \vec{F}_{\beta,\text{phyche},i} \qquad (5.2.1)$$

where

$\vec{F}_{\beta,i}$:   inter-particle force vector with the contact angle β at the i-th contact point,

$\vec{F}_{\beta,\text{grav},i}$:   inter-particle force vector due to self-weight (body force) with the contact angle β at the i-th contact point,

$\vec{F}_{\beta,\text{seep},i}$:   inter-particle force vector due to seepage force with the contact angle β at the i-th contact point,

$\vec{F}_{\beta,\text{matr},i}$:   inter-particle force vector due to surface tension with the contact angle β at the i-th contact point,

$\vec{F}_{\beta,\text{ext},i}$:   inter-particle force vector due to external force (surface force) with the contact angle β at the i-th contact point,

$\vec{F}_{\beta,\text{osmo},i}$:   inter-particle force vector due to osmotic pressure with the contact angle β at the i-th contact point and

$\vec{F}_{\beta,\text{phyche},i}$: inter-particle force vector due to physicochemical force with the contact angle β at the i-th contact point.

The inter-particle force vector due to physicochemical force $\vec{F}_{\beta,\text{phyche},i}$ related to the van der Waals force, electrostatic force and so on can be neglected for the coarse-grained soil. Additionally, as the pore water in the coarse-grained soil is generally considered not to have ionic and/or chemical components, the inter-particle force vector due to osmotic pressure $\vec{F}_{\beta,\text{osmo},i}$ is also neglected. Equation (5.2.1) is then rewritten as follows:

$$\vec{F}_{\beta,i} = \vec{F}_{\beta,\text{grav},i} + \vec{F}_{\beta,\text{seep},i} + \vec{F}_{\beta,\text{matr},i} + \vec{F}_{\beta,\text{ext},i} \qquad (5.2.2)$$

5.2.2 Inter-particle stress vector

The inter-particle stress vector $\vec{F}_\beta$ is defined as the average of resultant inter-particle force vectors per unit area on the plane with the inclination angle β. And then the following equation is obtained, referring to Equations (2.4.2) and (4.2.8):

$$\vec{F}_\beta = \sum_{i=1}^{N_{ca,\beta}} \vec{F}_{\beta,i} \tag{5.2.3}$$

where $N_{ca,\beta}$: number of contact points with contact angle β per unit area

Similar to Equation (5.2.2), the inter-particle stress vector can be expressed by the following equation:

$$\vec{F}_\beta = \vec{F}_{\beta,grav} + \vec{F}_{\beta,seep} + \vec{F}_{\beta,matr} + \vec{F}_{\beta,ext} \tag{5.2.4}$$

Each term in the right side of Equation (5.2.4) is related to the inter-particle force vectors as follows:

$$\vec{F}_{\beta,grav} = \sum_{i=1}^{N_{ca,\beta}} \vec{F}_{\beta,grav,i} \tag{5.2.5}$$

$$\vec{F}_{\beta,seep} = \sum_{i=1}^{N_{ca,\beta}} \vec{F}_{\beta,seep,i} \tag{5.2.6}$$

$$\vec{F}_{\beta,matr} = \sum_{i=1}^{N_{ca,\beta}} \vec{F}_{\beta,matr,i} \tag{5.2.7}$$

$$\vec{F}_{\beta,ext} = \sum_{i=1}^{N_{ca,\beta}} \vec{F}_{\beta,ext,i} \tag{5.2.8}$$

where

$N_{ca,\beta}$:   number of contact points with contact angle β per unit area,

$\vec{F}_\beta$:   inter-particle stress vector on the plane with the inclination angle β,

$\vec{F}_{\beta,grav}$: inter-particle stress vector due to gravitational force on the plane with the inclination angle β,

$\vec{F}_{\beta,seep}$: inter-particle stress vector due to seepage force on the plane with the inclination angle β,

$\vec{F}_{\beta,matr}$: inter-particle stress vector due to surface tension on the plane with the inclination angle β and

$\vec{F}_{\beta,ext}$: inter-particle stress vector due to external force on the plane with the inclination angle β.

Referring to Equation (2.4.8), the number of contact points with contact angle β per unit area, $N_{\text{ca},\beta}$, can be obtained as follows:

$$N_{\text{ca},\beta} = N_{\text{ca}} \cdot f_\beta(\beta) \Delta\beta \qquad (5.2.9)$$

where

N_{ca}:   total number of contact points per unit area (see Equation (4.3.7)) and

$f_\beta(\beta)$:   probability density function of contact angle (see Equation (3.2.7)).

The direction of an inter-particle stress vector due to gravity force $\vec{F}_{\beta,\text{grav}}$ expressed by Equation (5.2.5) is always vertical. The inter-particle stress vector due to seepage force $\vec{F}_{\beta,\text{seep}}$ expressed by Equation (5.2.6) is generated between two points in a soil block where the piezometric heads are different and its direction is normal to equipotential line (see Figure 5.10). The direction of inter-particle stress vector due to matric suction $\vec{F}_{\beta,\text{matr}}$ expressed by Equation (5.2.7) is always normal to the contact plane and then $\vec{F}_{\beta,\text{matr}}$ does not have the tangential component. The inter-particle stress vector due to external force $\vec{F}_{\beta,\text{ext}}$ expressed by Equation (5.2.8) corresponds to conventional stress, such as the axial stress, radial stress and so on generated in the triaxial specimen.

5.3 INTER-PARTICLE FORCE VECTOR AND INTER-PARTICLE STRESS VECTOR DUE TO GRAVITATIONAL FORCE

The inter-particle force vector and inter-particle stress vector due to gravitational force under dry conditions and hydrostatic conditions (saturated conditions) are considered in this section.

The direction of the inter-particle force vector and inter-particle stress vector due to gravitational force is always vertical and fixed. In the following, the gravitational acceleration is denoted by "g" which is used as both the vector and scalar quantities, although "g" is exactly the vector quantity.

5.3.1 Inter-particle force vector and inter-particle stress vector under dry conditions

Let's consider the cube with the unit length as shown in Figure 4.8. The relation between the inter-particle force vector and inter-particle stress vector with the contact angle β is expressed as follows:

$$\vec{F}_{\beta,\text{grav}} = \sum_{i=1}^{N_{\text{ca},\beta}} \vec{F}_{\beta,\text{grav},i} \qquad (5.2.5\text{bis})$$

where

$N_{ca,\beta}$: number of contact points with contact angle β per unit area (see Equation (5.2.9)),

$\vec{F}_{\beta,grav,i}$: inter-particle force vector at the i-th contact point with contact angle β, whose direction is vertical and

$\vec{F}_{\beta,grav}$: inter-particle stress vector due to gravitational force with contact angle β, whose direction is vertical.

The inter-particle stress vector $\vec{F}_{grav}$ due to gravitational force is related to the inter-particle stress vector with the contact angle β as follows, taking account of the resultant force per unit area acted on the plane.

$$\vec{F}_{grav} = \sum_{\beta=-\frac{\pi}{2}}^{\frac{\pi}{2}} \left(\frac{\vec{F}_{\beta,grav}}{\cos\beta} \right) \tag{5.3.1}$$

The following equation is obtained, referring to Equations (5.2.5) and (5.3.1).

$$\vec{F}_{\beta,grav} = \frac{N_{ca,\beta}}{N_{ca}} \cdot \vec{F}_{grav} \cdot \cos\beta \tag{5.3.2}$$

Substituting Equation (5.2.9) into Equation (5.3.2), $\vec{F}_{\beta,grav}$ is rewritten as follows:

$$\vec{F}_{\beta,grav} = \vec{F}_{grav} \cdot \cos\beta \cdot f_\beta(\beta)\Delta\beta \tag{5.3.3}$$

On the other hand, the inter-particle stress vector in the soil block with the unit cross-section area and the thickness D_{cha} shown in Figure 4.8 is expressed as follows, using the dry density ρ_d and the characteristic length D_{cha}:

$$\vec{F}_{grav} = \rho_d \cdot g \cdot D_{cha} \tag{5.3.4}$$

where

$\vec{F}_{grav}$: inter-particle stress vector due to gravitational force whose direction is vertical,

ρ_d: dry density of soil and

D_{cha}: characteristic length.

Equation (5.3.4) is interpreted as the equation which relates the gravitational force (body force) to the stress (surface force) in the soil block.

Substituting Equation (3.1.10) into Equation (5.3.4), the following equation is obtained:

$$\vec{F}_{grav} = \frac{1}{1+e} \cdot \rho_s \cdot g \cdot D_{cha} \tag{5.3.5}$$

where ρ_s: density of soil particle.

Substituting Equations (5.3.5) into Equation (5.3.3), the following equation is obtained:

$$\vec{F}_{\beta,grav} = \frac{1}{1+e} \cdot \rho_s \cdot g \cdot D_{cha} \cdot \cos\beta \cdot f_\beta(\beta)\Delta\beta \tag{5.3.6}$$

Then the normal and tangential components of inter-particle stress vector $\vec{F}_{\beta,grav}$ are obtained from Equation (5.3.6) as follows:

$$F_{\beta,grav,N} = \frac{1}{1+e} \cdot \rho_s \cdot g \cdot D_{cha} \cdot \cos^2\beta \cdot f_\beta(\beta)\Delta\beta \tag{5.3.7}$$

$$F_{\beta,grav,T} = \frac{1}{1+e} \cdot \rho_s \cdot g \cdot D_{cha} \cdot \sin\beta \cdot \cos\beta \cdot f_\beta(\beta)\Delta\beta \tag{5.3.8}$$

where

$F_{\beta,grav,N}$:   normal component (scalar quantity) of an inter-particle stress vector on the plane with the contact angle β due to gravitational force whose direction is vertical and

$F_{\beta,grav,T}$:   tangential component (scalar quantity) of an inter-particle stress vector on the plane with the contact angle β due to gravitational force whose direction is vertical.

For example, the normal and tangential components of an inter-particle stress vector due to gravitational force with contact angle, $\beta=0$, are obtained from Equations (5.3.7) and (5.3.8) as follows, which means the force acted on the horizontal plane.

$$F_{\beta=0,grav,N} = \frac{1}{1+e} \cdot \rho_s \cdot g \cdot D_{cha} \cdot f_\beta(0)\Delta\beta \tag{5.3.9}$$

$$\vec{F}_{\beta=0,grav,T} = 0 \tag{5.3.10}$$

It is found from Equations (5.3.9) and (5.3.10) that the inter-particle stress vector due to gravitational force acting on the horizontal plane has only a tangential component.

5.3.2 Inter-particle stress vector and pore water pressure under hydrostatic conditions

Firstly, we review Archimedes' principle before discussing the inter-particle stress vector and pore water pressure due to the gravitational force, because this principle plays an important role in deriving the inter-particle stress and pore water pressure acted on the plane under hydrostatic condition.

5.3.2.1 Archimedes' principle

Let's image the cubic container shown in Figure 5.2 which is filled with water and includes a spherical particle with the radius r. In the two-dimensional

Cartesian coordinates where the z-axis is vertical, the spherical particle can be expressed as follows:

$$(x - x_0)^2 + (z - z_0)^2 = r^2 \tag{5.3.11}$$

where

(x_0, z_0):   center point of spherical particle,
r: radius of spherical particle and
α: angle shown in Figure 5.2.

If the water pressure on a point (x,z) of circumference is expressed by $u_w(x,z)$, the force acted on the infinitesimal circumference $r d\alpha$ due to water pressure is $u_w(x,z) \, r d\alpha$, and then the vertical and horizontal components are obtained as follows:

Vertical component $u_w(x,z) \cdot r \cdot d\alpha \cdot \cos\alpha$ (5.3.12)

Horizontal component: $u_w(x,z) \cdot r \cdot d\alpha \cdot \sin\alpha$ (5.3.13)

where

$$x = x_0 + r \sin\alpha \tag{5.3.14}$$

$$z = z_0 - r \cos\alpha \tag{5.3.15}$$

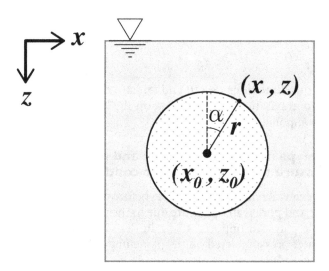

Figure 5.2 A sphere in the hydrostatic water.

The vertical and horizontal components of force acted upon the whole circumference due to water pressure are then obtained as follows:

$$\text{Vertical component: } \int_{\alpha=0}^{2\pi} u_w(x,z) \cdot r \cdot \cos\alpha \cdot d\alpha \tag{5.3.16}$$

$$\text{Horizontal component: } \int_{\alpha=0}^{2\pi} u_w(x,z) \cdot r \cdot \sin\alpha \cdot d\alpha \tag{5.3.17}$$

When the water is in the hydrostatic condition, the water pressure is expressed as follows:

$$u_w(x,z) = \rho_w \cdot g \cdot z \tag{5.3.18}$$

where

$u_w(x,z)$:   hydrostatic pressure,
ρ_w: density of water and
g: gravitational acceleration.

Substituting Equation (5.3.18) into Equations (5.3.16) and (5.3.17), the following equations are obtained:

$$\text{Vertical component: } \int_{\alpha=0}^{2\pi} \rho_w \cdot g \cdot z \cdot r \cdot \cos\alpha \cdot d\alpha$$

$$= \int_{\alpha=0}^{2\pi} \rho_w \cdot g \cdot (z_0 - r\cos\alpha) \cdot r \cdot \cos\alpha \cdot d\alpha$$

$$= \rho_w \cdot g \cdot z_0 \cdot r \int_{\alpha=0}^{2\pi} \cos\alpha \cdot d\alpha - \rho_w \cdot g \cdot r^2 \int_{\alpha=0}^{2\pi} \cos^2\alpha \, d\alpha \tag{5.3.19}$$

$$= \rho_w \cdot g \cdot z_0 \cdot r \left[\sin\alpha\right]_0^{2\pi} - \rho_w \cdot g \cdot r^2 \int_{\alpha=0}^{2\pi} \frac{1+\cos 2\alpha}{2} \, d\alpha$$

$$= -\rho_w \cdot g \cdot r^2 \left[\frac{1}{2}\alpha - \frac{\sin 2\alpha}{4}\right]_0^{2\pi}$$

$$= -\rho_w \cdot g \cdot \pi \cdot r^2$$

$$\text{Horizontal component: } \int_{\alpha=0}^{2\pi} \rho_w \cdot g \cdot z \cdot r \cdot \sin\alpha \cdot d\alpha = 0 \tag{5.3.20}$$

Equation (5.3.19) indicates that a spherical particle in the hydrostatic water is subjected to the upward force equal to the weight replaced a spherical particle by water, i.e., Archimedes' principle. On the other hand, Equation (5.3.20) means that the horizontal force acting on a spherical particle is not generated in the hydrostatic water. The pore water pressure expressed by Equation (5.3.18) is negative for the unsaturated soil, but Archimedes' principle is still applicable, i.e., the buoyant force is generally existed under the unsaturated condition, as well as the saturated condition.

5.3.2.2 Inter-particle stress vector and pore water pressure

Figure 5.3 shows a container and a spherical particle on the platform scale, where a particle is out of a container in Figure 5.3(a) and included in a container in Figure 5.3(b). The amounts shown by the indicator of platform scale $F_{\text{indicator}}$ are same and expressed as follows:

$$\text{For Figure 5.3(a):} \quad F_{\text{indicator}} = (M_s + M_w)g \tag{5.3.21}$$

$$\text{For Figure 5.3(b):} \quad F_{\text{indicator}} = \left\{ \left(M_s - \rho_w \frac{M_s}{\rho_s} \right) + \left(\frac{M_s}{\rho_s} + \frac{M_w}{\rho_w} \right) \rho_w \right\} g \tag{5.3.22}$$

$$= (M_s + M_w)g$$

where
  M_s: mass of spherical particle and
  M_w: mass of water in the container.

  Then, for Figure 5.3(b), the average force per unit area vertically acting on the bottom plane of the container due to gravitational force (i.e., pressure p_{bottom}) is expressed as follows:

$$p_{\text{bottom}} = \frac{F_{\text{indicator}}}{A} \tag{5.3.23}$$

where A: area of the bottom plane.

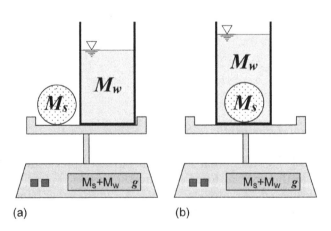

Figure 5.3 Weight of particle, water and container (a) Particle out of a container, (b) Particle in a container.

Substituting Equation (5.3.22) into Equation (5.3.23), the following equation is obtained:

$$p_{\text{bottom}} = \frac{M_s - \rho_w \dfrac{M_s}{\rho_s}}{A} g + \frac{M_w + \rho_w \dfrac{M_s}{\rho_s}}{A} g \qquad (5.3.24a)$$

Equation (5.3.24a) can be interpreted similarly to Figure 4.3 as follows:

$$\begin{aligned}(\text{pressure on the bottom plane}) &= (\text{inter} - \text{particle stress}) \\ &+ (\text{water pressure})\end{aligned} \qquad (5.3.24b)$$

The first term in Equation (5.3.24a) is rewritten as follows for the unit cube, referring to Figure 4.1.

$$\frac{M_s - \rho_w \dfrac{M_s}{\rho_s}}{A} g = \frac{\dfrac{1}{1+e} \cdot \rho_s - \rho_w \cdot V_s}{1 \times 1} g = \frac{\dfrac{1}{1+e} \cdot \rho_s - \dfrac{1}{1+e} \cdot \rho_w}{1 \times 1} g$$

$$= \frac{1}{1+e}(\rho_s - \rho_w) \cdot g \qquad (5.3.25)$$

Equation (5.3.25) can be extended to apply to the inter-particle stress vector for the soil block with the thickness D_{cha} in the hydrostatic saturated condition. And then the following equation instead of Equation (5.3.5) is obtained under hydrostatic condition:

$$\vec{F}_{\text{grav}} = \frac{1}{1+e} \cdot (\rho_s - \rho_w) \cdot g \cdot D_{\text{cha}} \qquad (5.3.26)$$

The inter-particle stress vector on the plane with the contact angle β is obtained as follows, as well as Equation (5.3.6):

$$\vec{F}_{\beta,\text{grav}} = \frac{1}{1+e} \cdot (\rho_s - \rho_w) \cdot g \cdot D_{\text{cha}} \cdot \cos\beta \cdot f_\beta(\beta) \Delta\beta \qquad (5.3.27)$$

The normal and tangential components of an inter-particle stress vector are expressed as follows, using Equation (5.3.27).

$$F_{\beta,\text{grav},N} = \frac{1}{1+e} \cdot (\rho_s - \rho_w) \cdot g \cdot D_{\text{cha}} \cdot \cos^2\beta \cdot f_\beta(\beta) \Delta\beta \qquad (5.3.28)$$

$$F_{\beta,\text{grav},T} = \frac{1}{1+e} \cdot (\rho_s - \rho_w) \cdot g \cdot D_{\text{cha}} \cdot \sin\beta \cdot \cos\beta \cdot f_\beta(\beta) \Delta\beta \qquad (5.3.29)$$

The second term in Equation (5.3.24) is the force due to water pressure which is then rewritten as follows for the unit cube, referring to Figure 4.1.

$$\frac{M_w + \rho_w \dfrac{M_s}{\rho_s}}{A} g = \frac{\dfrac{e\rho_w}{1+e} + \rho_w V_s}{1 \times 1} g = \frac{\dfrac{e\rho_w}{1+e} + \dfrac{1}{1+e}\rho_w}{1 \times 1} g = \rho_w g \qquad (5.3.30)$$

Therefore, the water pressure vertically acted on the bottom plane with the height h under hydrostatic condition is obtained as follows:

$$p_{w,\text{stat}} = \rho_w g \cdot h \qquad (5.3.31)$$

where $p_{w,\text{stat}}$: hydrostatic pressure acted on the bottom plane.

The total pressure of inter-particle stress vector and hydrostatic pressure vertically acting on the bottom plane under the hydrostatic saturated condition is expressed as follows, using Equations (5.3.26) and (5.3.31):

$$p_{\text{bottom}} = \vec{F}_{\text{grav}} + p_{w,\text{stat}} = \frac{\rho_s + e\rho_w}{1+e} g \cdot h \qquad (5.3.32)$$

where p_{bottom}: total pressure acted on the bottom plane, corresponding to the overburdened pressure of saturated soil block with the height h.

We can extend Equation (5.3.32) to the unsaturated condition, introducing the degree of saturation S_r, then Equation (5.3.32) is modified for the unsaturated soil as follows:

$$p_{\text{bottom}} = \vec{F}_{\text{grav}} + p_{w,\text{stat}} = \frac{\rho_s + eS_r\rho_w}{1+e} g \cdot h \qquad (5.3.33)$$

where S_r: degree of saturation.

Figure 5.4 shows the cubic sealed container with the height h which has a particle and water. When the extra negative pressure p_{suction} is given through the upper valve, Equation (5.3.32) is rewritten as follows:

$$p_{\text{bottom}} = \vec{F}_{\text{grav}} + p_{w,\text{stat}} + p_{\text{suction}} = \frac{\rho_s + e\rho_w}{1+e} g \cdot h + p_{\text{suction}} \qquad (5.3.34)$$

where h: height of container in Figure 5.4.

When the absolute amount of negative pressure p_{suction} is small, p_{bottom} is positive, which means the direction of p_{bottom} is downward, as shown in Figure 5.4(a). On the other hand, when the absolute amount of negative pressure p_{suction} is large, p_{bottom} becomes negative, which means the direction of p_{bottom} is upward, as shown in Figure 5.4(b).

The pressure in the vertical plane of container p_{side} is obtained as follows, because the inter-particle stress vector $\vec{F}_{\text{grav}}$ due to gravitational force is eliminated:

$$p_{\text{side}} = p_{w,\text{stat}} + p_{\text{suction}} = \frac{e\rho_w}{1+e} g \cdot h + p_{\text{suction}} \qquad (5.3.35)$$

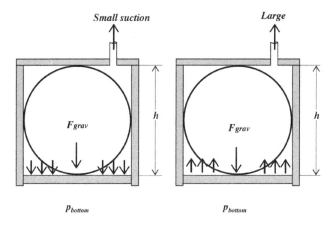

Figure 5.4 Water pressure acted on the bottom plate in sealed container.

The extra negative pressure $p_{suction}$ in Equation (5.3.35) corresponds to the negative pore air pressure for the vacuum consolidation and the pore water pressure for the pumping test. The total pressures p_{bottom} and p_{side} acting on the horizontal and vertical plane respectively will be used when the bearing capacity and earth pressure are discussed in Chapter 8.

5.4 INTER-PARTICLE FORCE VECTOR AND INTER-PARTICLE STRESS VECTOR DUE TO SEEPAGE FORCE

5.4.1 Bernoulli's principle

Bernoulli (1700–1782) derived the equation of motion for the steady flow of inviscid fluid called Bernoulli's principle in 1738. In this sub-section, Bernoulli's principle is reviewed before discussing inter-particle force and inter-particle stress due to seepage force because the piezometric head in Bernoulli's principle plays a key role with respect to the flow of water in saturated–unsaturated soil.

Figure 5.5 shows the circular straight pipe with the minute cross-section area dA and the minute length ds in the direction of x-axis, where the x-axis has an angle θ to the horizontal axis and the z-axis is taken as the vertical axis. The pressure $p(s)$ and $p(s+ds) = p + (dp/ds)ds$ are applied the both ends of pipe as shown in Figure 5.5. The weight of fluid in the minute pipe is calculated to be $\rho_f \cdot ds \cdot dA \cdot g$, where ρ_f is the density of inviscid fluid, and g is the acceleration of gravity.

Figure 5.6 shows the change in velocity along the flow line and then the acceleration a_t is obtained as follows:

$$a_t = \frac{v(s+ds)-v(s)}{dt} = \frac{v(s)+\dfrac{dv}{ds}\cdot vdt -v(s)}{dt} = v\frac{dv}{ds} \tag{5.4.1}$$

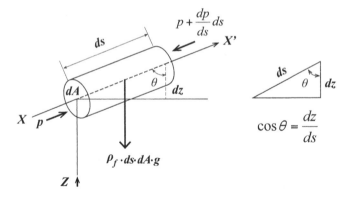

Figure 5.5 Force equilibrium of inviscid fluid in minute pipe.

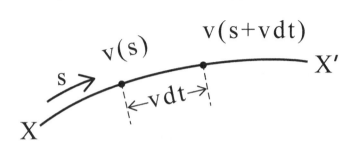

Figure 5.6 Change in velocity along flow line.

The force equilibrium for the minute pipe in the direction of x-axis is then obtained as follows:

$$p(s)dA - p(s+ds)dA - \rho_f \cdot ds \cdot dA \cdot g \cdot \cos\theta = \rho_f \cdot ds \cdot dA \cdot \left(v\frac{dv}{ds}\right) \quad (5.4.2)$$

where

$$p(s) - p(s+ds) = -\frac{dp}{ds}ds,$$

$$\cos\theta = \frac{dz}{ds}$$

Then Equation (5.4.2) is rewritten as follows, dividing both sides of Equation (5.4.2) by $\rho_f \cdot ds \cdot dA$:

$$\frac{1}{\rho_f}\frac{dp}{ds} + g \cdot \frac{dz}{ds} + v\frac{dv}{ds} = 0 \quad (5.4.3)$$

Furthermore Equation (5.4.3) is rewritten as follows:

$$\frac{1}{\rho_f} dp + g \cdot dz + v \cdot dv = 0 \tag{5.4.4}$$

Integrating Equation (5.4.4), the following equation can be obtained:

$$\int \frac{1}{\rho_f} dp + g \cdot z + \int v \cdot dv = \text{const.} \tag{5.4.5}$$

Equation (5.4.5) is rewritten as follows for the inviscid and incompressible fluid with the density ρ_w:

$$p + \rho_w \cdot g \cdot z + \frac{1}{2} \rho_w v^2 = P = \text{const.} \tag{5.4.6}$$

where P: total pressure.

Equation (5.4.6) is generally called Bernoulli's principle for inviscid and incompressible fluid with respect to the pressure.

Introducing the concept of water head, Equation (5.4.6) is rewritten as follows:

$$h_{\text{pre}} + h_{\text{ele}} + h_{\text{vel}} = h_{\text{total}} \tag{5.4.7}$$

where

$h_{\text{pre}} = \dfrac{p}{\rho_w g}$:   pressure head,

$h_{\text{ele}} = z$:   elevation head,

$h_{\text{vel}} = \dfrac{v^2}{2g}$:   velocity head and

$h_{\text{total}} = \dfrac{P}{\rho_w g}$:   total head.

Equation (5.4.7) is called Bernoulli's principle for inviscid and incompressible fluid with respect to the water head. As the velocity head is infinitesimal through soil, the piezometric head defined in the following equation is often used for the seepage behavior in soil:

$$h_{\text{piez}} = h_{\text{pre}} + h_{\text{ele}} \tag{5.4.8}$$

where h_{piez}: piezometric head.

Figure 5.7 shows the water flow in the horizontal circular pipe with different diameters, where the datum is the center of pipe. Then the following equations are obtained, applying Bernoulli's principle to the water level in the side tubes, using Equation (5.4.7):

$$h_{\text{total}} = H \tag{5.4.9}$$

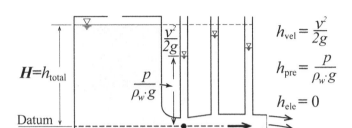

Figure 5.7 Water flow in horizontal pipe with different diameter.

$$h_{pre} = \frac{p}{\rho_w g} \qquad (5.4.10)$$

$$h_{vel} = \frac{v^2}{2g} \qquad (5.4.11)$$

$$h_{ele} = 0 \qquad (5.4.12)$$

It is difficult for the real groundwater flow to identify the height H in Figure 5.7. Additionally, the velocity head of groundwater is much smaller compared with the pressure head. Then the piezometric head h_{piez} expressed by Equation (5.4.8) is generally used to analyze the groundwater flow instead of the total head, i.e., the distribution of h_{piez} expressed by the elevation is essential and important to analyze the groundwater flow.

Figure 5.8(a) and (b) show the hydrostatic and hydrodynamic conditions of horizontal pipe with side tubes respectively, where only water is included in the horizontal pipe with side tubes. Under the hydrostatic condition, the water levels in the side tubes are the same as the height of water tank

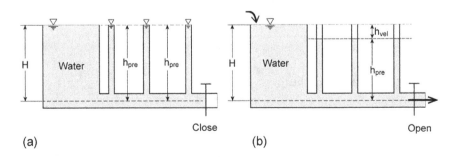

(a)

(b)

Figure 5.8 Water level in side tubes for inviscid fluid (a) Hydrostatic condition, (b) Hydrodynamic condition.

H at three points as shown in Figure 5.8(a), i.e., the following equation is obtained:

$$H = h_{pre} \tag{5.4.13}$$

Under the hydrodynamic condition, the water levels are also the same at three points, but the water levels in the side tubes are different from the height of the water tank H, because the velocity head h_{vel} is generated by the flow in the pipe, i.e., the following equation is obtained:

$$H = h_{pre} + h_{vel} \tag{5.4.14}$$

For the viscous fluid, the frictional head loss must additionally be taken into account.

Figure 5.9(a) and (b) also show the hydrostatic and hydrodynamic conditions as well as Figure 5.8(a) and (b), except that the saturated soil is included in the horizontal pipe. In other words, the water filling the pipe in Figures 5.7 and 5.8 is replaced by the saturated soil (=soil particles + pore water) in Figure 5.9. Under the hydrostatic condition, the water levels in the side tube are the same at three points, as shown in Figure 5.9(a). On the other hand, under the hydrodynamic condition, the water level in the side tube which corresponds to the piezometric head is proportionally decreased to zero at the right end, as shown in Figure 5.9(b), i.e., the piezometric head h_{piez} at the right end is the same as the elevation head h_{ele}. It is considered that the difference in the cross-sectional area of horizontal pipe in Figure 5.7 corresponds to the difference in void ratio in Figure 5.9(b).

Figure 5.10 shows the flow line and the equipotential line in the pipe filled with the saturated soil, which is the same condition as Figure 5.9(b). The equipotential line is orthogonal to the flow line under the steady condition. Additionally, we assume that this relation between the equipotential line and flow line can be applied under the unsteady condition in this book. In Figure 5.10, the water level in each side tube expresses the pressure head h_{ele} which corresponds to the hydrostatic pressure, and the inclination lined

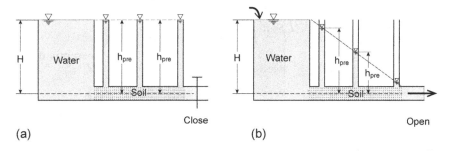

Figure 5.9 Water level in side tubes for soil (a) Hydrostatic condition, (b) Hydrodynamic condition.

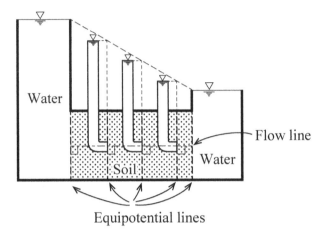

Figure 5.10 Equipotential line and flow line.

with the water level in side tubes expresses the hydraulic gradient, which relates to the hydrodynamic pressure which is discussed in Section 5.4.2.

Here let's consider the vertical flow in soil, using Figures 5.11 and 5.12.

Figure 5.11(a), (b) and (c) show the hydrostatic and hydrodynamic conditions of containers filled with water, saturated soil and unsaturated soil respectively.

In Figure 5.11(a) and (b), the piezometric heads h_{piez} at the top and bottom of container are same, i.e., the following equations are obtained when the datum is the bottom of container.

$$h_{piez,top} = h_{pre} + h_{ele} = 0 + h = h \tag{5.4.15}$$

$$h_{piez,bottom} = h_{pre} + h_{ele} = h + 0 = h \tag{5.4.16}$$

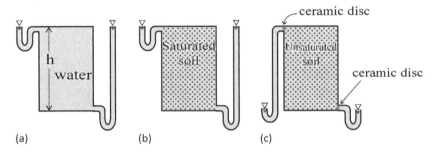

Figure 5.11 Hydrostatic conditions of containers filled with water, saturated soil and unsaturated soil (a) Container filled water, (b) Container filled with saturated soil, and (c) Container filled with unsaturated soil.

where

$h_{\text{piez,top}}$:   piezometric head at the top of container in Figure 5.11(a) and (b) and

$h_{\text{piez,bottom}}$:   piezometric head at the bottom of container in Figure 5.11(a) and (b).

Then the flow does not occur under hydrostatic condition because the piezometric heads are the same at the top and bottom of the container, as shown in Equations (5.4.15) and (5.4.16).

Figure 5.11(c) shows the container filled with unsaturated soil, where the piezometric head h_{piez} at the top and bottom of the container is located at the bottom of container. As the piezometric head at the top and bottom are the same, the pore water in the unsaturated soil does not flow as well as in Figure 5.11(a) and (b). However, the pore water pressures at the top and bottom are negative and zero respectively, i.e., the following equations are obtained when the datum is the bottom of container:

$$h_{\text{piez,top}} = h_{\text{pre}} + h_{\text{ele}} = -h + h = 0 \qquad (5.4.17)$$

$$h_{\text{piez,bottom}} = h_{\text{pre}} + h_{\text{ele}} = 0 + 0 = 0 \qquad (5.4.18)$$

where

$h_{\text{piez,top}}$:   piezometric head at the top of container in Figure 5.11(c) and
$h_{\text{piez,bottom}}$:   piezometric head at the bottom of container in Figure 5.11(c).

Figure 5.12(a) shows the water flow in the container filled with water, where the flow velocity is $v = \sqrt{2g \cdot h}$. Figure 5.12(b) shows the water flow in the container filled with saturated soil. Applying Darcy's law, the flow

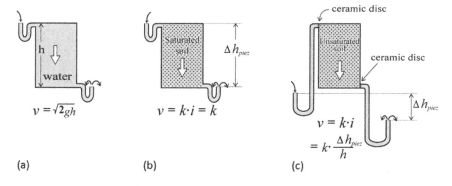

Figure 5.12 Hydrodynamic conditions of containers filled with water, saturated soil and unsaturated soil (a) Container filled with water, (b) Container filled with saturated soil, and (c) Container filled with unsaturated soil.

velocity is $v = k \cdot i$, where k is the saturated permeability coefficient and i is the hydraulic gradient. In Figure 5.12(b), the hydraulic gradient is 1 and the velocity v is equal to the saturated permeability coefficient k. Figure 5.12(c) shows the unsaturated condition, where the piezometric head h_{piez} at the top the of container is located lower than the bottom of the container, but higher than that at the bottom of container. Applying Darcy's law, the flow velocity is $v = k \cdot i$, where k is the unsaturated permeability coefficient and i is the hydraulic gradient defined as $i = \Delta h_{piez} / h$.

It is found from Figures 5.10 and 5.11 that the flow of pore water in both saturated and unsaturated soil depends on the piezometric head. That is to say, when the difference in piezometric head between two points is zero, the pore water is under the hydrostatic condition and does not flow. On the other hand, when the difference in piezometric head between two points is not zero, the pore water is under the hydrodynamic condition and the pore water flows in soil irrespective of saturated and unsaturated conditions. It means that the piezometric head is one of the most important physical quantities for the pore water in soil.

5.4.2 Seepage force

5.4.2.1 Definition

Figure 5.13(a) shows the water flow through the saturated soil in which the different piezometric heads are given at the right and left ends. As the piezometric head at the left end is higher than the right end, the water flows from left to right. The piezometric heads at both ends, $h_{piez}(x_1)$ and $h_{piez}(x_2)$, are expressed as follows:

$$h_{piez}(x_1) = z_1 + \frac{u_1}{\rho_w g} = h_{ele,1} + h_{pre,1} \tag{5.4.19}$$

$$h_{piez}(x_2) = z_2 + \frac{u_2}{\rho_w g} = h_{ele,2} + h_{pre,2} \tag{5.4.20}$$

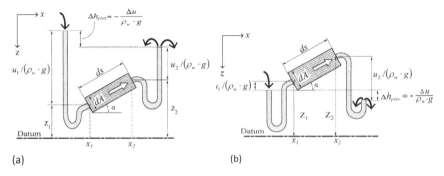

(a)   (b)

Figure 5.13 Water flow in soil (a) Saturated Soil, (b) Unsaturated Soil.

The difference in the pressure between both ends, Δu, is obtained from Equations (5.4.19) and (5.4.20) as follows:

$$\Delta u = \rho_w g \left(h_{\text{piez}}(x_1) - h_{\text{piez}}(x_2) \right) = \rho_w g \cdot \Delta h_{\text{piez}} \tag{5.4.21}$$

where Δh_{piez}: difference of the piezometric head.

The seepage force in the saturated soil, J_{sat}, acting on both ends, can be obtained from the following equation, referring to Figure 5.13(a) and Equation (5.4.21):

$$J_{\text{sat}} = \Delta u \cdot dA = \rho_w g \cdot \Delta h_{\text{piez}} \cdot dA \tag{5.4.22}$$

where dA: cross-section area of pipe in Figure 5.13(a).

Then the seepage force per unit volume, j_{sat}, is obtained from Equation (5.4.22) as follows:

$$j_{\text{sat}} = \frac{J_{\text{sat}}}{dV} = \frac{J_{\text{sat}}}{dA \cdot ds} = \rho_w g \cdot \frac{\Delta h_{\text{piez}}}{ds} = \rho_w g \cdot i \tag{5.4.23}$$

where
  ds: length of pipe in Figure 5.13(a) and
  $i = \Delta h_{\text{piez}}/ds$: hydraulic gradient.

Equation (5.4.23) is the equation which defines the seepage force per unit volume for saturated soil. Additionally, it is found that the seepage force per unit volume, j_{sat}, is classified into body force as well as gravitational force.

5.4.2.2 Simple example

Figure 5.14 shows a sphere with diameter D_s in the cubic container where the water flows in the vertical direction under the hydraulic gradient $i = \Delta h_{\text{pre}}/D_s$. In this condition, the seepage force is obtained as follows, referring to Equations (5.4.22) and (5.4.23).

$$J_{\text{sat}} = j_{\text{sat}} \cdot V = \rho_w g \cdot i \cdot \left\{ \frac{\pi}{6} D_s^3 + \left(1 - \frac{\pi}{6} \right) D_s^3 \right\} \tag{5.4.24}$$

Where
  V: volume of container,

$$i = \frac{\Delta h_{\text{pre}}}{D_s}$$

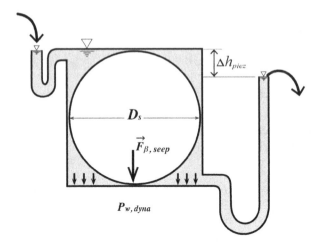

Figure 5.14 Inter-particle force and water pressure on the bottom plate.

Then the seepage force per unit cube acting on the bottom plane of container is obtained as follows, referring to the phase diagram shown in Figure 4.1:

$$j_{\text{sat}} = \rho_w g \left\{ \frac{\pi}{6} + \left(1 - \frac{\pi}{6}\right) \right\} \cdot i = \frac{1}{1+e}\rho_w g \cdot i + \frac{e}{1+e}\rho_w g \cdot i = \rho_w g \cdot i \quad (5.4.25)$$

It is found from Equation (5.4.25) that Equation (5.4.23) is applicable to the special case shown in Figure 5.14, and the seepage force j_{sat} consists of two components which are related to the inter-particle stress vector and water pressure.

When $\Delta h_{\text{piez}} = D_s$ in Figure 5.14, Equation (5.4.24) is rewritten as follows:

$$J_{\text{sat}} = \rho_w g \cdot \Delta h_{\text{pre}} \cdot D_s^2, \quad (5.4.26)$$

then the pressure acted on the bottom plane of unit cube is obtained from Equation (5.4.26) as follows:

$$p_{\text{bottom}} = \frac{J_{\text{sat}}}{D_s^2} = \rho_w g \cdot \Delta h_{\text{pre}} = \rho_w g \cdot D_s \quad (5.4.27)$$

It is found from Equation (5.4.27) that the seepage force per unit volume due to the vertical flow shown in Figure 5.14 is same as the hydrostatic pressure when $i = 1$.

Figure 5.13(b) shows the water flow through the unsaturated soil in which the different piezometric heads are given at the right and left ends for unsaturated soil. As the piezometric heads at both ends are different, the water flows from left to right just as in Figure 5.13(a). In this condition, the seepage force per unit cube j_{sat} is obtained to extend Equation (5.4.25) as follows:

$$j_{unsat} = \frac{1 + eS_r}{1 + e} \rho_w g \cdot i \tag{5.4.28}$$

where S_r: degree of saturation.

It may be concluded in this subsection that 1. the hydrostatic pressure and hydrodynamic pressure are independent each other; 2. the hydrostatic pressure relates to the buoyance force (see Section 5.3.2); and 3. the hydrodynamic pressure relates to the seepage force which acts on the plane normal to the flow line, i.e., tangential to the equipotential line.

5.4.3 Inter-particle force vector and inter-particle stress vector under hydrodynamic conditions

The direction of seepage force is same as that of the water flow direction, i.e., the direction of the hydraulic gradient. Both the seepage force and gravitational force are body forces which relate to mass of body. The direction of seepage force is determined by the normal line to the contour lines of piezometric head, although the gravitational force is always vertical.

The inter-particle stress vector due to seepage force, $\vec{F}_{\beta,seep}$, acting on the plane with the angle β is related to the inter-particle force vector $\vec{F}_{\beta,seep,i}$ as follows:

$$\vec{F}_{\beta,seep} = \sum_{i=1}^{N_{ca,\beta}} \vec{F}_{\beta,seep,i} \tag{5.2.6 bis}$$

The inter-particle stress vector $\vec{F}_{seep}$ due to seepage force is related to the inter-particle stress vector $\vec{F}_{\beta,seep}$ as follows, similar to Equation (5.3.1):

$$\vec{F}_{seep} = \sum_{\beta = -\frac{\pi}{2}}^{\frac{\pi}{2}} \left(\frac{\vec{F}_{\beta,seep}}{\cos(\theta_{ave} - \beta)} \right) \tag{5.4.29}$$

where θ_{ave}: angle between the vertical direction and the direction normal to predominant pore water flow, i.e., the direction normal to the equipotential line, as shown in Figure 5.15.

The following equation is obtained, referring to Equations (5.2.6) and (5.4.29):

$$\vec{F}_{\beta,seep} = \frac{N_{ca,\beta}}{N_{ca}} \cdot \vec{F}_{seep} \cdot \cos(\theta_{ave} - \beta) \tag{5.4.30}$$

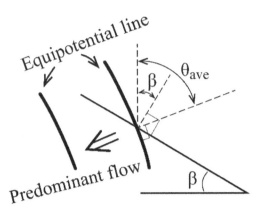

Figure 5.15 Relation between β and θ_{ave}.

Substituting Equation (5.2.9) into Equation (5.4.30), $\vec{F}_{\beta,\text{seep}}$ is rewritten as follows:

$$\vec{F}_{\beta,\text{seep}} = \vec{F}_{\text{seep}} \cdot \cos(\theta_{\text{ave}} - \beta) \cdot f_\beta(\beta) \Delta\beta \tag{5.4.31}$$

On the other hand, the inter-particle stress vector $\vec{F}_{\text{seep}}$ can be expressed as follows, referring to Equations (5.3.5) and (5.4.23):

$$\vec{F}_{\text{seep}} = \frac{1}{1+e} \rho_w g \cdot i \cdot D_{\text{cha}} = \frac{1}{1+e} \rho_w g \cdot \Delta h_{\text{piez}} \tag{5.4.32}$$

where

$i = \left(\Delta h_{\text{piez}} / D_{\text{cha}} \right)$:   hydraulic gradient determined by the contours of piezometric head and

Δh_{piez}:   difference of piezometric head between the thickness D_{cha}.

Substituting Equation (5.4.32) into Equation (5.4.31), the following equation is obtained:

$$\vec{F}_{\beta,\text{seep}} = \frac{1}{1+e} \rho_w g \cdot \Delta h_{\text{piez}} \cdot \cos(\theta_{\text{ave}} - \beta) \cdot f_\beta(\beta) \Delta\beta \tag{5.4.33}$$

The normal and tangential components of the inter-particle stress vector due to seepage force are obtained as follows:

$$F_{\beta,\text{seep},N} = \frac{1}{1+e} \rho_w g \cdot \Delta h_{\text{piez}} \cdot \cos^2(\theta_{\text{ave}} - \beta) \cdot f_\beta(\beta) \Delta\beta \tag{5.4.34}$$

$$F_{\beta,\text{seep},T} = \frac{1}{1+e} \rho_w g \cdot \Delta h_{\text{piez}} \cdot \sin(\theta_{\text{ave}} - \beta) \cdot \cos(\theta_{\text{ave}} - \beta) \cdot f_\beta(\beta) \Delta\beta \tag{5.4.35}$$

The hydrodynamic pressure acting on the plane normal to the water flow direction is expressed as follows, referring to Figure 5.14 where $D_s = D_{cha}$:

$$p_{w,\text{dyna}} = \frac{e}{1+e}\rho_w g \cdot i \cdot D_{cha} = \frac{e}{1+e}\rho_w g \cdot \Delta h_{\text{piez}} \qquad (5.4.36)$$

where $p_{w.\text{dyna}}$: hydrodynamic pressure acting on the plane normal to the water flow direction.

The total pressure p_{bottom} acting on the bottom plane is obtained as follows, referring to Equations (5.4.34) and (5.4.36), because of $\beta = \theta_{ave} = 0$:

$$p_{\text{bottom}} = F_{\beta=0,\text{seep},N} + p_{w,\text{dyna}} = \rho_w g \cdot \Delta h_{\text{piez}} \qquad (5.4.37)$$

Comparing Equation (5.4.37) with (5.4.27), it is found that the total pressure acting on the bottom plane obtained from Equation (5.4.37) is the same as the pressure obtained from Equation (5.4.27) with respect to the seepage force, if $D_s = D_{cha} = \Delta h_{\text{piez}}$ in Figure 5.14, i.e., $i = 1$.

Under unsaturated conditions, Equation (5.4.36) is modified as follows, referring to Equation (5.4.28):

$$p_{w,\text{dyna}} = \frac{eS_r}{1+e}\rho_w g \cdot i \cdot D_{cha} = \frac{eS_r}{1+e}\rho_w g \cdot \Delta h_{\text{piez}} \qquad (5.4.38)$$

The total pressure p_{bottom} acting on the bottom plane is obtained as follows under the unsaturated condition:

$$p_{\text{bottom}} = F_{\beta=0,\text{seep},N} + p_{w,\text{dyna}} = \frac{1+eS_r}{1+e} \cdot \rho_w g \cdot \Delta h_{\text{piez}} \qquad (5.4.39)$$

Comparing Equation (5.4.38) with $p_{w,\text{stat}}$ in Equation (5.3.33), it is found that the inter-particle stress vector due to seepage force with the hydraulic gradient $i = -1$, i.e., $\Delta h_{\text{piez}} = D_{cha}$, is same as the water pressure acting on the bottom plane in the saturated–unsaturated condition.

In the case where the direction of water flow through the saturated soil is vertical ($\theta_{ave} = 0$) and upward, the equilibrium for the seepage and gravity forces on the horizontal plane ($\beta = 0$) of unit cube is obtained as follows, referring to Equations (5.3.26) and (5.4.25):

$$\frac{1}{1+e}(\rho_s - \rho_w)g = \rho_w g \cdot i_{cri} \qquad (5.4.40)$$

where i_{cri}: critical hydraulic gradient.

Then the following equation is obtained from Equation (5.4.40):

$$i_{cri} = \frac{\rho_s/\rho_w - 1}{1+e} = \frac{G_s - 1}{1+e} \qquad (5.4.41)$$

Equation (5.4.41) is the popular equation which defines the critical hydraulic gradient for quicksand in the conventional soil mechanics.

5.5 INTER-PARTICLE FORCE VECTOR AND INTER-PARTICLE STRESS VECTOR DUE TO SURFACE TENSION

5.5.1 Capillary rise and suction due to surface tension

Figure 5.16 shows a narrow glass tube inserted into water under atmospheric conditions. The water level in the glass tube is higher than the free water level due to the surface tension of water. When the free water surface is taken as the datum and the upward direction of z-axis is positive, as shown in Figure 5.16, the following equation is obtained, referring to Equation (5.4.8):

$$h_{ele} + h_{pre} = h_{piez} = 0 \tag{5.5.1}$$

where
  h_{ele}: elevation head,
  h_{pre}: pressure head and
  h_{piez}: piezometric head.

As the elevation head in the glass tube higher than the free water is proportionally increased to positive, the pressure head in Equation (5.5.1) has to be decreased to negative as shown in Figure 5.16, i.e., the water pressure is less than atmospheric pressure.

The force equilibrium of the water in the vertical direction of the narrow glass tube can be expressed as follows:

$$\pi \cdot d \cdot T_s \cdot \cos\alpha = \rho_w \cdot g \cdot \frac{\pi}{4} \cdot d^2 \cdot h_{ca} \tag{5.5.2}$$

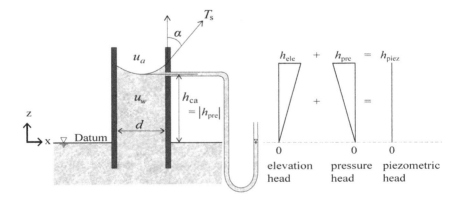

Figure 5.16 Capillary rise in the narrow glass tube.

where

  T_s: surface tension,
  d: diameter of capillary tube,
  α: contact angle between glass wall and water,
  ρ_w: density of water,
  g: gravitational acceleration and
  h_{ca}: capillary height.

Equation (5.5.2) is written as follows:

$$h_{ca} = \frac{4T_s \cos\alpha}{\rho_w g \cdot d} \tag{5.5.3}$$

And h_{ca} is equal to the absolute value of the negative pressure head h_{pre} in Equation (5.5.1).

The following equation can be derived by considering the force equilibrium at the interface between water and air in the tube in Figure 5.16.

$$u_a - u_w = \rho_w g \cdot h_{ca} \tag{5.5.4}$$

where

  u_a: air pressure (= atmospheric pressure) and
  u_w: water pressure at the interface of glass tube.

Substituting Equation (5.5.3) into Equation (5.5.4), the following equation is obtained:

$$u_a - u_w = \frac{4T_s \cos\alpha}{d} \tag{5.5.5}$$

As $u_a - u_w$ in Equation (5.5.5) is defined as the matric suction s_u in the unsaturated soil mechanics, Equation (5.5.5) is rewritten as follows:

$$s_u = u_a - u_w = \frac{4T_s \cos\alpha}{d} \tag{5.5.6}$$

5.5.2 Two-particles' model

The inter-particle force due to surface tension is usually generated at a contact point of adjacent soil particles which are irregular in size and shape, as shown in Figure 3.2. Because it is difficult to evaluate qualitatively the irregularity in size and shape of real soil particles, we assume that the inter-particle force due to surface tension can be obtained from two adjacent spheres with same radius r kept in the water at the contact point, as shown in Figure 5.17 which is a special case of two adjacent spheres shown in Figure 3.9(b). The water formed at a contact point is kept by the meniscus due to surface tension and the direction of inter-particle force is normal to

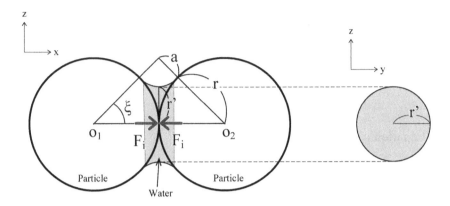

Figure 5.17 Two-particles' model.

the contact plane, which is called the "two-particles' model" in the following discussion. The force due to surface tension F_i at a contact point is then
obtained as follows:

$$F_i = 2\pi r' \cdot T_s + \pi r'^2 \cdot s_u \qquad (5.5.7)$$

where

F_i: inter-particle force at the contact point,
T_s: surface tension,
r': radius shown in Figure 5.17 and
s_u: suction defined by Equation (5.5.6).

Figure 5.18 shows the curvature of liquid membrane under the pressure
p_0 and p acting on the top and back sides respectively. Then the following
equation is obtained:

$$p - p_0 = T_s \cdot \left(\frac{1}{R_1} + \frac{1}{R_2} \right) \qquad (5.5.8)$$

The pressures p_0 and p in Figure 5.18 correspond to u_w and u_a in Figure 5.17
respectively. The radii R_1 and R_2 correspond to a and $-r'$ respectively.
Equation (5.5.8) is then rewritten as follows:

$$u_a - u_w = T_s \cdot \left(\frac{1}{a} - \frac{1}{r'} \right) \qquad (5.5.9)$$

Equation (5.5.8) is also rewritten as follows:

$$s_u = T_s \cdot \left(\frac{1}{a} - \frac{1}{r'} \right) \qquad (5.5.10)$$

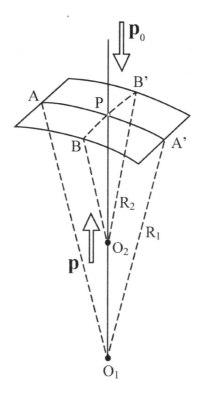

Figure 5.18 Curvature of liquid membrane and pressure.

Furthermore, the following equations are obtained from the geometrical relation of 2 spheres with same radius shown in Figure 5.17:

$$\tan\xi = \frac{a+r'}{r} \tag{5.5.11}$$

$$\cos\xi = \frac{r}{a+r} \tag{5.5.12}$$

As the variables T_s, s_u and r are known, the unknown variables r', a and ξ can be calculated from Equations (5.5.10), (5.5.11) and (5.5.12). Consequently r' is expressed as follows, using T_s, s_u and r:

$$r' = \frac{-3T_s + \sqrt{9T_s^2 + 8r \cdot s_u \cdot T_s}}{2s_u} \tag{5.5.13}$$

Equation (5.5.13) is rewritten as follows, using the diameter D_s instead of the radius r.

$$r' = \frac{-3T_s + \sqrt{9T_s^2 + 4D_s \cdot s_u \cdot T_s}}{2s_u} \tag{5.5.14}$$

5.5.3 Inter-particle force vector and inter-particle stress vector derived from the two-particles' model

The direction of the inter-particle force vector due to surface tension at a contact point is limited to be normal to the tangential plane. The inter-particle force vector due to surface tension at the i-th contact point is then expressed as follows, referring to Equation (5.1.1):

$$\vec{F}_{\beta,\text{matr},i} = F_{\beta,\text{matr},i,N} \cdot \vec{e}_{\beta,\text{matr},i,N} \tag{5.5.15}$$

Using Equation (5.5.7), the following equation is obtained, as the inter-particle force vector does not depend the contact angle:

$$F_{\beta,\text{matr},i,N} = 2\pi r' \cdot T_s + \pi r'^2 \cdot s_u \tag{5.5.16}$$

It has been shown that the inter-particle stress vector is expressed as follows:

$$\vec{F}_{\beta,\text{matr}} = \sum_{i=1}^{N_{\text{ca},\beta}} \vec{F}_{\beta,\text{matr},i} \tag{5.2.7bis}$$

Substituting Equation (5.5.16) into Equation (5.2.7), the following equation is obtained:

$$\vec{F}_{\beta,\text{matr}} = \sum_{i=1}^{N_{\text{ca},\beta}} F_{\beta,\text{matr},i,N} \cdot \vec{e}_{\beta,\text{matr},i,N} \tag{5.5.17}$$

Substituting Equation (5.5.16) into Equation (5.5.17), the following equation is obtained:

$$\vec{F}_{\beta,\text{matr}} = \sum_{i=1}^{N_{\text{ca},\beta}} \left(2\pi r'_i \cdot T_s + \pi r_i'^2 \cdot s_u \right) \cdot \vec{e}_{\beta,\text{matr},i,N}$$

$$= 2\pi \cdot T_s \sum_{i=1}^{N_{\text{ca},\beta}} r'_i \cdot \vec{e}_{\beta,\text{matr},i,N} + \pi \cdot s_u \sum_{i=1}^{N_{\text{ca},\beta}} r_i'^2 \cdot \vec{e}_{\beta,\text{matr},i,N} \tag{5.5.18}$$

Therefore, the following equation is obtained:

$$F_{\beta,\text{matr},N} = 2\pi \cdot T_s \sum_{i=1}^{N_{\text{ca},\beta}} r'_i + \pi \cdot s_u \sum_{i=1}^{N_{\text{ca},\beta}} r_i'^2 \tag{5.5.19}$$

where $N_{\text{ca},\beta}$: number of contact points with contact angle β per unit area and obtained by Equation (5.2.9).

Referring to Equation (2.4.2), Equation (5.5.19) is rewritten as follows:

$$F_{\beta,\text{matr},N} = N_{\text{ca},\beta} \cdot \left(2\pi \cdot T_s \cdot E[r'_i] + \pi \cdot s_u \cdot E[r_i'^2] \right) \tag{5.5.20}$$

Referring to Equation (2.4.7), $E[r_i']$ and $E[r_i'^2]$ are expressed as follows:

$$E[r_i'] = \int_0^\infty r' \cdot f_s(D_s) dD_s \qquad (5.5.21)$$

$$E[r_i'^2] = \int_0^\infty r'^2 \cdot f_s(D_s) dD_s \qquad (5.5.22)$$

Substituting Equation (5.5.14) into Equations (5.5.21) and (5.5.22), the following equations are obtained:

$$E[r_i'] = \int_0^\infty \frac{-3T_s + \sqrt{9T_s + 4D_s \cdot s_u \cdot T_s}}{2s_u} \cdot f_s(D_s) dD_s \qquad (5.5.23)$$

$$E[r_i'^2] = \int_0^\infty \left(\frac{-3T_s + \sqrt{9T_s + 4D_s \cdot s_u \cdot T_s}}{2s_u} \right)^2 \cdot f_s(D_s) dD_s \qquad (5.5.24)$$

It is concluded from Equations (5.5.20), (5.5.23), and (5.5.24) that the normal component of inter-particle stress vector $F_{\beta,matr,N}$ can be obtained from the grain size distribution $f_s(D_s)$ (see Equation (3.2.6)), number of contact points $N_{ca,\beta}$ (see Equation (4.3.7)), surface tension of water T_s and suction s_u.

5.6 INTER-PARTICLE FORCE VECTOR AND INTER-PARTICLE STRESS VECTOR DUE TO EXTERNAL FORCE

5.6.1 Mohr's stress circle

Mohr's stress circle has been used in conventional soil mechanics, as it is convenient to analyze the mechanical behaviors under the axisymmetric stress condition. Mohr's stress circle is also an important tool for analyzing the inter-particle force and the inter-particle stress due to the external force in our approach and is briefly reviewed in this subsection.

Figure 5.19 shows a stress state generated in a small triangle element. The compressional stress and anti-clockwise shear stress are taken to be positive. The equations for the force equilibrium in the x- and z-axes are shown as follows:

For x-axis $\quad \sigma_x dz + \tau_{zx} dx - \sigma_{\beta,N} \sin\beta ds + \tau_\beta \cos\beta ds = 0 \qquad (5.6.1)$

For z-axis $\quad -\sigma_z dx - \tau_{xz} dz + \sigma_{\beta,N} \cos\beta ds + \tau_\beta \sin\beta ds = 0 \qquad (5.6.2)$

where $dx = ds \cos\beta$ and $dz = ds \sin\beta$

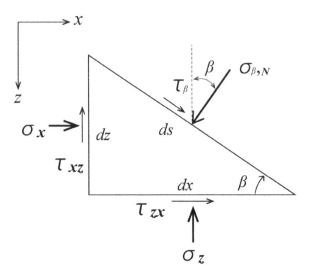

Figure 5.19 Equilibrium of small element.

As the resultant momentum is zero for the small element in the equilibrium condition, the following equation is derived:

$$\tau_{zx} = \tau_{xz} \tag{5.6.3}$$

The following equations can be obtained from Equations (5.6.1), (5.6.2) and (5.6.3):

$$\sigma_{\beta,N} = \sigma_x \sin^2 \beta + \sigma_z \cos^2 \beta + 2\tau_{zx} \sin\beta\cos\beta \tag{5.6.4}$$

$$\tau_\beta = (\sigma_z - \sigma_x)\sin\beta\cos\beta + \tau_{zx}\left(\sin^2 \beta - \cos^2 \beta\right) \tag{5.6.5}$$

When the directions of maximum and minimum principal stresses σ_1 and σ_3 coincide with the z- and x-axes respectively, as shown in Figure 5.20 (i.e., x-axis is horizontal and z-axis is vertical), τ_{zx} becomes zero. The following equation is then obtained:

$$\sigma_{\beta,N} = \sigma_1 \cos^2 \beta + \sigma_3 \sin^2 \beta = \frac{\sigma_1 + \sigma_3}{2} + \frac{\sigma_1 - \sigma_3}{2}\cos 2\beta \tag{5.6.6}$$

$$\tau_\beta = (\sigma_1 - \sigma_3)\sin\beta\cos\beta = \frac{\sigma_1 - \sigma_3}{2}\sin 2\beta \tag{5.6.7}$$

where

$$\tan\phi = \frac{\tau_\beta}{\sigma_{\beta,N}} \tag{5.6.8}$$

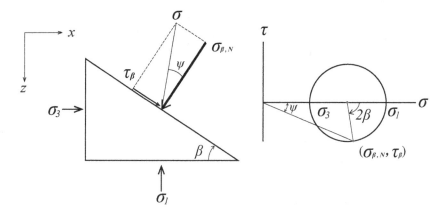

Figure 5.20 Relation between principal stresses and stresses on the plane with angle β.

The normal and tangential stresses on the plane with the inclination angle β is obtained from Equations (5.6.6) and (5.6.7) and expressed as the point $(\sigma_{\beta,N}, \tau_\beta)$ on the Mohr's circle. Figure 5.21(a) and (b) shows a stress state in the triaxial specimen and the corresponding stress point on the Mohr's stress circle respectively. It is found from Figure 5.21 that the stress state on a given plane with the inclination angle β in the triaxial specimen can be obtained through the Mohr's stress circle if the amount and direction of principal stresses σ_1 and σ_3 are given.

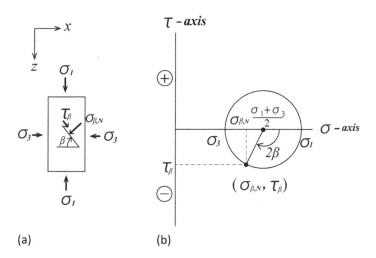

(a) (b)

Figure 5.21 Stress state in triaxial specimen and Mohr's stress circle (a) Stress state in triaxial specimen, (b) Mohr's stress circle.

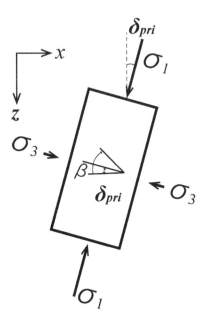

Figure 5.22 Stress state where the direction of maximum principal stress is inclined to be δ_{pri} from vertical.

Figure 5.22 shows the small element where the direction of maximum principal stress σ_1 is different from the vertical direction by δ_{pri}. In this condition, Equations (5.6.6) and (5.6.7) are rewritten as follows:

$$\sigma_{\beta,N} = \sigma_1 \cos^2\left(\beta - \delta_{\text{pri}}\right) + \sigma_3 \sin^2\left(\beta - \delta_{\text{pri}}\right)$$

$$= \frac{\sigma_1 + \sigma_3}{2} + \frac{\sigma_1 - \sigma_3}{2}\cos 2\left(\beta - \delta_{\text{pri}}\right) \tag{5.6.9}$$

$$\tau_\beta = \left(\sigma_1 - \sigma_3\right)\sin\left(\beta - \delta_{\text{pri}}\right)\cos\left(\beta - \delta_{\text{pri}}\right)$$

$$= \frac{\sigma_1 - \sigma_3}{2}\sin 2\left(\beta - \delta_{\text{pri}}\right) \tag{5.6.10}$$

where δ_{pri}: direction angle between vertical and maximum principal stress.

5.6.2 Inter-particle force vector and inter-particle stress vector derived from Mohr's stress circle

Figure 5.23 shows the i-th contact point taken from the triaxial specimen where the contact angle is β and the inter-particle force vector is $\vec{F}_{\beta,\text{ext},i}$. We can then obtain the following relation between the inter-particle force vector $\vec{F}_{\beta,\text{ext},i}$ and the inter-particle stress vector $\vec{F}_{\beta,\text{ext}}$ as follows:

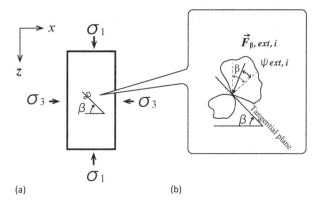

Figure 5.23 Inter-particle force at the *i*-th contact point with the contact angle β taken from triaxial specimen (a) A contact point with contact angle β in the triaxial specimen under stress condition of σ_1 and σ_3, (b) Inter-particle force at the *i*-th contact point due to external force.

$$\vec{F}_{\beta,\text{ext}} = \sum_{i=1}^{N_{\text{ca},\beta}} \vec{F}_{\beta,\text{ext},i} \qquad (5.2.8\text{bis})$$

The inter-particle stress vector due to external force is assumed to be the stress used in the conventional soil mechanics and the following equation is then obtained:

$$F_{\beta,\text{ext},N} = \sigma_{\beta,N} \qquad (5.6.11)$$

$$F_{\beta,\text{ext},T} = \tau_\beta \qquad (5.6.12)$$

where

$\quad F_{\beta,\text{ext},N}$: normal component of inter-particle stress vector, scalar quantity,
$\quad F_{\beta,\text{ext},T}$: tangential component of inter-particle stress vector, scalar quantity,
$\quad \sigma_{\beta,N}$: normal stress generated on the plane with the contact angle β and
$\quad \tau_\beta$: tangential stress generated on the plane with the contact angle β.

Referring to Figure 5.23(b) and Equation (2.4.2), the following equation is obtained:

$$E\left[\psi_{\beta,\text{ext},i}\right] = \tan^{-1}\frac{\tau_\beta}{\sigma_{\beta,N}} \qquad (5.6.13)$$

where $E\left[\psi_{\beta,\text{ext},i}\right]$: mean value of $\psi_{\beta,\text{ext},i}$.

It is found from the discussion in this section that the inter-particle stress vector due to external force is obtained if the direction and amount of principal stresses σ_1 and σ_3 used in the conventional soil mechanics are known.

5.7 SUMMARY FOR NORMAL AND TANGENTIAL COMPONENTS OF INTER-PARTICLE STRESS VECTOR

In this section, we summarize the results for the inter-particle stress vector and water pressure because the potential slip plane and the self-weight retaining height discussed in Chapter 8 are deeply related to the ratio of tangential to normal inter-particle stress vector, $F_{\beta,T}/F_{\beta,N}$.

The inter-particle stress vector for coarse-grained soil is expressed as follows:

$$\vec{F}_\beta = \vec{F}_{\beta,\text{grav}} + \vec{F}_{\beta,\text{seep}} + \vec{F}_{\beta,\text{matr}} + \vec{F}_{\beta,\text{ext}} \tag{5.2.4bis}$$

Using the normal and tangential components of each inter-particle stress vector, Equation (5.2.4) is rewritten as follows:

$$F_{\beta,N} = F_{\beta,\text{grav},N} + F_{\beta,\text{seep},N} + F_{\beta,\text{matr},N} + F_{\beta,\text{ext},N} \tag{5.7.1}$$

$$F_{\beta,T} = F_{\beta,\text{grav},T} + F_{\beta,\text{seep},T} + F_{\beta,\text{ext},T} \tag{5.7.2}$$

The normal and tangential components of inter-particle stress vectors due to gravitational force are expressed as follows, where the direction of gravitational force is vertical:

For dry soil:

$$F_{\beta,\text{grav},N} = \frac{1}{1+e} \cdot \rho_s \cdot g \cdot D_{cha} \cdot \cos^2 \beta \cdot f_\beta(\beta) \Delta\beta \tag{5.3.7bis}$$

$$F_{\beta,\text{grav},T} = \frac{1}{1+e} \cdot \rho_s \cdot g \cdot D_{cha} \cdot \sin\beta \cdot \cos\beta \cdot f_\beta(\beta) \cdot \beta \tag{5.3.8bis}$$

For saturated–unsaturated soil:

$$F_{\beta,\text{grav},N} = \frac{1}{1+e} \cdot (\rho_s - \rho_w) \cdot g \cdot D_{cha} \cdot \cos^2 \beta \cdot f_\beta(\beta) \Delta\beta \tag{5.3.28bis}$$

$$F_{\beta,\text{grav},T} = \frac{1}{1+e} \cdot (\rho_s - \rho_w) \cdot g \cdot D_{cha} \cdot \sin\beta \cdot \cos\beta \cdot f_\beta(\beta) \Delta\beta \tag{5.3.29bis}$$

Additionally, the total pressure acting on the horizontal and vertical planes under hydrostatic conditions is respectively expressed as follows:

$$p_{\text{bottom}} = \vec{F}_{\text{grav}} + p_{w,\text{stat}} + p_{\text{suction}} = \frac{\rho_s + eS_r\rho_w}{1+e} g \cdot h + p_{\text{suction}} \tag{5.3.34bis}$$

$$p_{\text{side}} = p_{w,\text{stat}} + p_{\text{suction}} = \frac{eS_r\rho_w}{1+e} g \cdot h + p_{\text{suction}} \tag{5.3.35bis}$$

The normal and tangential components of inter-particle stress vectors due to seepage force are expressed as follows, where the direction of seepage force is same as that of pore water flow:

$$F_{\beta,\text{seep},N} = \frac{1}{1+e}\rho_w g \cdot \Delta h_{\text{piez}} \cdot \cos^2\left(\theta_{\text{ave}} - \beta\right) \cdot f_\beta\left(\beta\right)\Delta\beta \qquad (5.4.34\text{bis})$$

$$F_{\beta,\text{seep},T} = \frac{1}{1+e}\rho_w g \cdot \Delta h_{\text{piez}} \cdot \sin\left(\theta_{\text{ave}} - \beta\right) \cdot \cos\left(\theta_{\text{ave}} - \beta\right) \cdot f_\beta\left(\beta\right)\Delta\beta \qquad (5.4.35\text{bis})$$

Additionally, the water pressure acting on the plane normal to the direction of average water flow is expressed as follows:

$$p_{w,\text{dyna}} = \frac{eS_r}{1+e}\rho_w g \cdot \Delta h_{\text{piez}} \qquad (5.4.36\text{bis})$$

The inter-particle stress vector due to matric suction has only the normal component and is expressed as follows:

$$F_{\beta,\text{matr},N} = N_{\text{ca},\beta} \cdot \left(2\pi \cdot T_s \cdot E[r_i'] + \pi \cdot s_u \cdot E[r_i'^2]\right) \qquad (5.5.20\text{bis})$$

The normal and tangential components of inter-particle stress vectors due to external force are expressed as follows, where $\sigma_{\beta,N}$ and τ_β can be calculated when the amount and direction of principal stresses are known:

$$F_{\beta,\text{ext},N} = \sigma_{\beta,N} \qquad (5.6.10\text{bis})$$

$$F_{\beta,\text{ext},T} = \tau_\beta \qquad (5.6.11\text{bis})$$

Chapter 6

Modeling of pore water retention by elementary particulate model (EPM)

In this chapter, we will microscopically analyze the water retention mechanism by means of the elementary particulate model (EPM) shown in Figure 3.8(c).

6.1 SOIL SUCTION

The term "soil suction" means the total suction which is the summation of matric suction and osmotic suction, i.e., the soil suction s_u which is expressed by the following equation:

$$s_u = s_{u,\text{total}} = s_{u,\text{matr}} + s_{u,\text{osmo}} \tag{6.1.1}$$

where s_u: soil suction, $s_{u,\text{total}}$: total suction, $s_{u,\text{matr}}$: matric suction and $s_{u,\text{osmo}}$: osmotic suction.

Osmotic suction can be neglected in coarse-grained soil, and then soil suction is expressed by the following equation:

$$s_u = s_{\text{total}} = s_{\text{matr}} \tag{6.1.2}$$

Then matric suction is abbreviated to be termed simply "suction," i.e., $s_u = s_{u,\text{matr}}$, in the following sections in this chapter. Matric suction due to the surface tension of water has been defined as the following equation in Section 5.5:

$$s_u = u_a - u_w = \frac{4T_s \cos \alpha}{d} \tag{5.5.6bis}$$

Applying Equation (5.5.6) to the EPM shown in Figure 3.9(c), the diameter of capillary tube d shown in Figure 5.16 is replaced by the diameter of pipe, D_v. The suction of EPM is then rewritten as follows:

$$s_u = u_a - u_w = \frac{4T_s \cos \alpha}{D_v} \tag{6.1.3}$$

6.2 MODELING OF SOIL WATER CHARACTERISTIC CURVE

The soil water characteristic curve is used as the collective term for suction–water content relation, suction–volumetric water content relation and suction–degree of saturation relation, where the suction is expressed by the normal scale and/or logarithmic scale. Furthermore, the soil water characteristic curves are classified into the main drying curve (MDC), main wetting curve (MWC) and scanning curves, as shown in Figure 6.1. The experimental data are often processed to obtain the physical quantities such as volumetric water content and degree of saturation, assuming that the volume change in the specimen does not occur when the suction changes, although the volume change really occurs with the change in suction. To analyze the water retention mechanism accurately, the problems for coupling between particulate soil structure and pore structure should be solved; this is discussed in Chapter 10.

The numerical calculation procedure to obtain one of the soil water characteristic curves, i.e., the suction–water content relation by means of the elementary particulate model, is discussed in this section.

If the grain size distribution and the void ratio are given for a soil type, the distribution parameters of pore size distribution, λ_v and ζ_v in

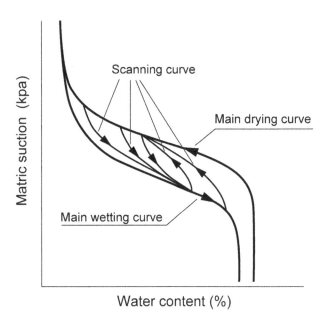

Figure 6.1 Hysteresis of soil water characteristic curve (After Editorial Committee for Japanese Geotechnical Society Standards, 2009 and modified).

Equation (3.2.8), can be obtained according to the numerical calculation procedure shown in Figure 3.13, using the following equations:

$$f_v(D_v) = \frac{1}{\sqrt{2\pi}\zeta_v \cdot D_v} \exp\left\{-\frac{(\ln D_v - \lambda_v)^2}{2\zeta_v^2}\right\} \qquad (3.2.8\text{bis})$$

$$e = \int_0^\infty \int_{-\frac{\pi}{2}}^{\frac{\pi}{2}} \frac{\varphi_{\text{EPM},p(D_v,\theta)}}{\varphi_{\text{EPM},s}(D_v,\theta)} \cdot f_v(D_v) \cdot f_c(\theta) d\theta dD_v \qquad (3.2.18\text{bis})$$

If a value of suction is given, the corresponding diameter D_v is back-calculated by using Equation (6.1.3), which is considered to be the maximum diameter d_w of the pipe filled with water in the elementary particulate model (DEM). Then the water content corresponding to the given suction is numerically calculated by the following equation, i.e., a plot of suction–water content relation can be obtained:

$$w = \frac{\rho_w}{\rho_s} \cdot \int_0^{d_w} \int_{-\frac{\pi}{2}}^{\frac{\pi}{2}} \frac{\varphi_{\text{EPM},p(D_v,\theta)}}{\varphi_{\text{EPM},s}(D_v,\theta)} \cdot f_v(D_v) \cdot f_c(\theta) d\theta dD_v \qquad (3.2.19\text{bis})$$

Figure 6.2 shows the calculation procedure to obtain the suction vs the water content when the void ratio does not change with the change in

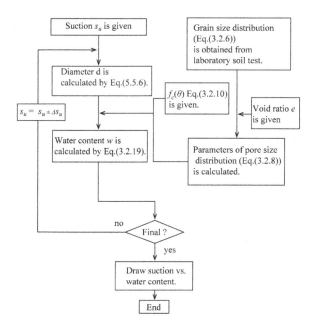

Figure 6.2 Calculation procedure for suction vs water content.

suction. In Figure 6.2 "$s_u = s_u + \Delta s_u$" and "$s_u = s_u - \Delta s_u$" correspond to the drying process and the wetting process respectively.

The other soil water characteristic curves (suction–degree of saturation and suction– volumetric water content) can be obtained by transforming the water content to volumetric water content and degree of saturation, which are respectively related to the water content by Equations (3.1.7) and (3.1.8) derived from the phase diagram shown in Figure 3.3.

6.3 MODELING OF HYSTERESIS OF SOIL WATER CHARACTERISTIC CURVE

6.3.1 Ink-bottle model

An ink-bottle tube shown in Figure 6.3 has been used to explain the hysteresis of the soil water characteristic curve. When the ink-bottle tube is inserted into the pad filled with water, as shown in Figure 6.3(a), the water level rises and stops at the height where the diameter of the tube changes from smaller to larger. On the other hand, when the ink-bottle tube filled with water is inserted into the pad, as shown in Figure 6.3(b), the water level falls and stops at the height where the diameter of tube changes from smaller to larger. The rising and falling of the water level in the ink-bottle tube correspond to the drying and wetting processes in the soil water characteristic curves respectively. Figure 6.3 is available to assist in understanding the hysteresis of the soil water characteristic curve conceptually.

In our approach, two kinds of the unit of ink-bottle are introduced for the drying and wetting processes of the soil water characteristic curve; the

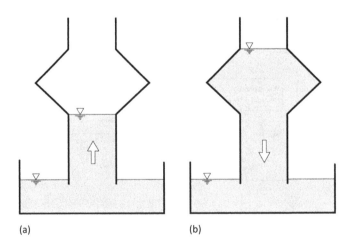

(a)   (b)

Figure 6.3 Ink-bottle model to explain capillary height for upward and downward of water level (a) Raising process, (b) Falling process.

barreled ink-bottle unit shown in Figure 6.4(a) is used for the drying process of the soil water characteristic curve and the nipped-in ink-bottle unit
shown in Figure 6.4(b) is used for the wetting process of the soil water characteristic curve. These ink-bottle units are called the ink-bottle model and
correspond to the pipe in the elementary particulate model (EPM) shown
in Figure 3.11. The barreled ink-bottle unit shown in Figure 6.4(a) is used
for the drying process and the nipped-in ink-bottle shown in Figure 6.4(b)
is used for the wetting process. The parameters prescribed the shape and
size of ink-bottle are the mean diameter d_{mean}, the height of ink-bottle h_{bo}
and the inclination angle β_{bo} as shown in Figure 6.4. The mean diameter
d_{mean} corresponds to the diameter of the glass tube shown in Figure 5.15.
Here the inclination angle α_{bo} is tentatively taken to be $45°(=\pi/4)$ and then
the height of ink-bottle h_{bo} is same as d_{mean} in the following:

$$h_{bo} = d_{mean} \tag{6.3.1}$$

Furthermore, the maximum and minimum diameters d_{max} and d_{min} are
obtained from the mean diameter d as follows:

$$d_{max} = \frac{5}{4} d_{mean} \tag{6.3.2}$$

$$d_{max} = \frac{3}{4} d_{mean} \tag{6.3.3}$$

Therefore, the volumes of both barreled and nipped-in ink-bottle unit, V_{unit},
can be obtained as follows:

$$V_{unit} = \frac{\pi}{4} \cdot d_{mean}^2 \cdot h_{bo} \tag{6.3.4}$$

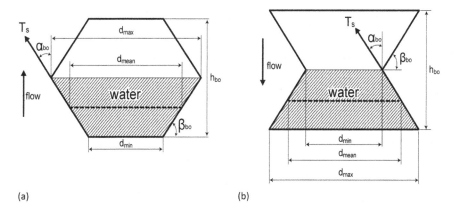

(a) (b)

Figure 6.4 Barreled ink-bottle unit and nipped-in ink-bottle unit (a) Barreled ink-bottoke
unit for drying process, (b) Nipped-in ink-bottle unit for wetting process.

Substituting Equation (6.3.1) into Equation (6.3.4), the following equation is obtained:

$$V_{unit} = \frac{\pi}{4} \cdot d_{mean}^3 \tag{6.3.5}$$

6.3.2 Main drying curve (MDC) and main wetting curve (MWC)

Figure 6.5(a) and (b) are the zigzag pipes consisting of piled barreled ink-bottle units and piled nipped-in ink-bottle units which are applied to the main drying curve (MDC) and main wetting curve (MWC) shown in Figure 6.1 respectively. It is assumed here that the water level of the unit matches the level of d_{max} for the drying process shown in Figure 6.4(a) and that the water level of the unit matches the level of d_{min} for the wetting process shown in Figure 6.4(b). Additionally, the contact angle α formed between the water and the glass wall shown in Figure 5.16 is also assumed to be zero. The equations of force equilibrium due to surface tension is then

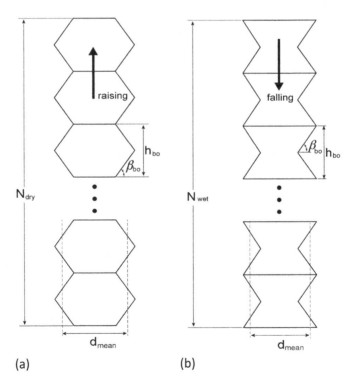

(a)   (b)

Figure 6.5 Ink-bottle model adopted in our approach (a) Barreled ink-bottle units, (b) Nipped-in ink-bottle units.

derived for the drying and wetting processes as follows, which correspond to Equation (5.5.2):

For the drying process:

$$\pi \cdot d_{max} \cdot T_s \cdot \cos\alpha_{bo} = \rho_w g \cdot V_{unit} \cdot \left(N_{dry} + \frac{1}{2}\right) \tag{6.3.6}$$

where N_{dry}: number of barreled ink-bottle units shown in Figure 6.5(a).

For the wetting process:

$$\pi \cdot d_{min} \cdot T_s \cdot \cos\alpha_{bo} = \rho_w g \cdot V_{unit} \cdot \left(N_{wet} + \frac{1}{2}\right) \tag{6.3.7}$$

where N_{wet}: number of nipped-in ink-bottle units shown in Figure 6.5(b).

The following equations are obtained from Equations (6.3.6) and (6.3.7) respectively.

$$N_{dry} \cong \frac{\pi \cdot d_{max} \cdot T_s \cdot \cos\alpha_{bo}}{\rho_w g \cdot V_{unit}} \tag{6.3.8}$$

$$N_{wet} \cong \frac{\pi \cdot d_{min} \cdot T_s \cdot \cos\alpha_{bo}}{\rho_w g \cdot V_{unit}} \tag{6.3.9}$$

When the inclination angle α_{bo} is tentatively taken to be 45° ($=\pi/4$), Equations (6.3.8) and (6.3.9) are rewritten as follows, using Equations (6.3.1), (6.3.2) and (6.3.3).

$$N_{dry} \cong \frac{5\pi \cdot T_s}{\sqrt{2} \cdot \rho_w g \cdot d_{mean}{}^2} \tag{6.3.10}$$

$$N_{wet} \cong \frac{3\pi \cdot T_s}{\sqrt{2} \cdot \rho_w g \cdot d_{mean}^2} \tag{6.3.11}$$

Figure 6.6 shows an example of the relation between d_{mean} and the numbers of units, N_{dry} and N_{wet}, calculated by Equations (6.3.10) and (6.3.11) for drying and wetting processes.

The following equations for suction are derived from Equations (6.3.10) and (6.3.11):

For the drying process:

$$s_u = \rho_w g \cdot d_{mean} \cdot N_{dry} = \frac{5\pi \cdot T_s}{\sqrt{2}d_{mean}} \tag{6.3.12}$$

For the wetting process:

$$s_u = \rho_w g \cdot d_{mean} \cdot N_{wet} = \frac{3\pi \cdot T_s}{\sqrt{2}d_{mean}} \tag{6.3.13}$$

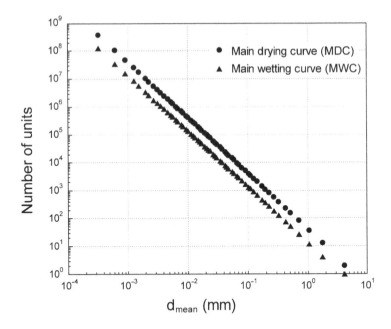

Figure 6.6 Relation between d, N_{dry} and N_{wet}.

6.3.3 Scanning drying curve (SDC) and scanning wetting curve (SWC)

Figure 6.7 shows schematic soil water characteristic curves, i.e., the main drying curve (MDC), the main wetting curve (MWC) and the scanning wetting curve (SWC). The water content and suction at points A, B, C, D, E and F in Figure 6.7 are respectively denoted by $A(w_0, s_{u,\text{MDC},0})$, $B(w_0, s_{u,\text{MWC},0})$, $C(w_0, s_{u,\text{SWC},0})$, $D(w_1, s_{u,\text{MDC},1})$, $E(w_1, s_{u,\text{MWC},1})$ and $F(w_1, s_{u,\text{SWC},1})$. In this subsection, the procedure to identify the unknown water content and suction of point F, i.e., $F(w_1, s_{u,\text{SWC},1})$, is shown under the conditions that $A(w_0, s_{u,\text{MDC},0})$, $B(w_0, s_{u,\text{MWC},0})$, $C(w_0, s_{u,\text{SWC},0})$, $D(w_1, s_{u,\text{MDC},1})$ and $E(w_1, s_{u,\text{MWC},1})$ are known.

Figure 6.8(a) shows the MDC and MWC in 3-D space where the water content w is separated into w_{dry}-axis and w_{wet}-axis which are rotated anti-clockwise and clockwise by 45°from the s_u–w plane respectively. Figure 6.8(b) shows the surface of s_u~w_{dry}~w_{wet} relation with the contour lines of water content which are drawn by rotating MEC and MWC on the s_u–w plane. Figure 6.9 shows the quadrant of contour lines projected on the w_{dry}~w_{wet} plane. The circles with the water content w_0 and w_1, are expressed as follows:

$$w_{\text{wet}}^2 + w_{\text{dry}}^2 = w_0^2 \qquad (6.3.14)$$

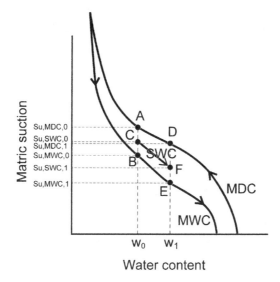

Figure 6.7 Scanning wetting curve (SWC), MDC and MWC.

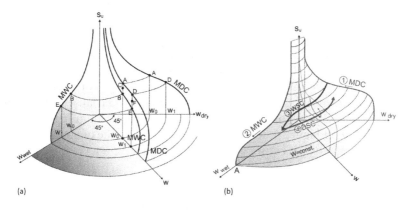

Figure 6.8 Transformation of soil water characteristic curves (a) Movement of MDS and NWC into 3D space, (b) Surface of soil-water characteristic curves.

$$w_{wet}^2 + w_{dry}^2 = w_1^2 \qquad (6.3.15)$$

Points A, B, C, D, E and F in Figure 6.9 correspond to those in Figures 6.7 and 6.8. Here we assume that $\overline{AC} : \overline{BC}$ in Figure 6.7 is same as $\widehat{AC} : \widehat{BC}$ in Figure 6.9 and then the following equation is obtained:

$$\left(s_{u,MDC,0} - s_{u,SWC,0}\right) : \left(s_{u,SWC,0} - s_{u,MWC,0}\right)$$

$$= \left(w_0 \cdot \alpha_o\right) : \left(w_0 \cdot \left(\frac{\pi}{2} - \alpha_o\right)\right) \qquad (6.3.16)$$

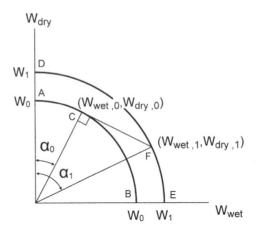

Figure 6.9 Quadrant contour of gravimetric water content.

The following equation is obtained from Equation (6.3.16):

$$\alpha_0 = \frac{\pi}{2} \cdot \frac{s_{u,\mathrm{MDC},0} - s_{u,\mathrm{SWC},0}}{s_{u,\mathrm{MDC},0} - s_{u,\mathrm{MWC},0}} \tag{6.3.17}$$

If the suction at point C, $s_{u,\mathrm{SWC},0}$, in Figure 6.7 is given, we can obtain α_0 from Equation (6.3.17) because $s_{u,\mathrm{MDC},0}$, $s_{u,\mathrm{MWC},0}$ are known.

Next, in Figure 6.9, we draw the tangential line of the quadrant circle from point C. The intersection point F is assumed to correspond to the point F on the scanning wetting curve in Figure 6.7, because the distance between point C and point F in Figure 6.9 is the minimum length on the surface shown in Figure 6.8(b). The tangential line of the quadrant of circle through the point C in Figure 6.9 is expressed as follows:

$$\sin \alpha_0 \cdot w_{\mathrm{wet}} + \cos \alpha_0 \cdot w_{\mathrm{dry}} = w_0 \tag{6.3.18}$$

Using Equations (6.3.15) and (6.3.18), the following equation is obtained with respect to w_{wet} at the point F:

$$w_{\mathrm{wet}}^2 - 2w_0 \cdot \sin \alpha_0 \cdot w_{\mathrm{wet}} + \left(w_0^2 - w_1^2 \cdot \sin^2 \alpha_0 \right) = 0 \tag{6.3.19}$$

The following equation is then obtained by solving Equation (6.3.16) with respect to w_{wet}:

$$w_{\mathrm{wet},1} = \sin \alpha_0 \cdot w_0 + \cos \alpha_0 \cdot \sqrt{w_1^2 - w_0^2} \tag{6.3.20}$$

Substituting Equation (6.3.20) into Equation (6.3.18), the following equation is obtained:

$$w_{\mathrm{dry},1} = w_0 - \tan \alpha_0 \cdot \left(\sin \alpha_0 \cdot w_0 + \cos \alpha_0 \cdot \sqrt{w_1^2 - w_0^2} \right) \tag{6.3.21}$$

Then α_1 in Figure 6.8 is obtained from the following equation, using Equations (4.3.20) and (4.3.21):

$$\tan\alpha_1 = \frac{w_{\text{dry}}}{w_{\text{wet}}} \qquad (6.3.22)$$

The following relation is also assumed, as well as Equation (6.3.16):

$$\left(s_{u,\text{MDC},1} - s_{u,\text{SWC},1}\right) : \left(s_{u,\text{SWC},1} - s_{u,\text{MWC},1}\right)$$
$$= \left(w_1 \cdot \alpha_1\right) : \left(w_1 \cdot \left(\frac{\pi}{2} - \alpha_1\right)\right) \qquad (6.3.23)$$

The following equation is obtained from Equation (6.3.23):

$$s_{u,\text{SWC},1} = \frac{2}{\pi}\left\{\alpha_1 \cdot s_{u,\text{MDC},1} + \left(\frac{\pi}{2} - \alpha_1\right)s_{u,\text{MWC},1}\right\} \qquad (6.3.24)$$

As α_1, $s_{u,\text{MDC},1}$ and $s_{u,\text{MWC},1}$ are known in Equation (6.3.24), $s_{u,\text{SWC},1}$ can be obtained.

The calculation procedure for the scanning wet curve is shown in Figure 6.10.

It is assumed that the scanning drying curve (SDC) is reversible when the water contents w_0 and w_1 are fixed, i.e., the path from point F to point C follows the same path as that from point C to point F without hysteresis. But, when the

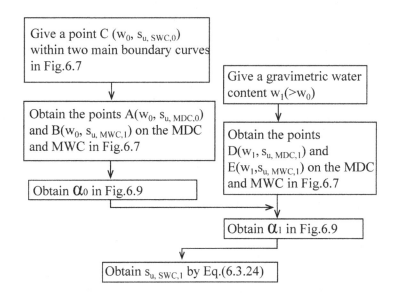

Figure 6.10 Calculation procedure for scanning wetting curve.

combinations of water contents $A(w_0, s_{u,\text{MDC},0})$, $B(w_0, s_{u,\text{MWC},0})$, $C(w_0, s_{u,\text{SWC},0})$, $D(w_1, s_{u,\text{MDC},1})$ and $E(w_1, s_{u,\text{MWC},1})$ in Figure 6.8 are changed, the new SWC and/or SDC are obtained according to the procedure shown in Figure 6.10.

6.4 CORRECTION OF PORE SIZE DISTRIBUTION FOR SOIL WATER CHARACTERISTIC CURVE

6.4.1 Correction method for soil water characteristic curve

The pore size distribution obtained by Equation (3.2.8) has been derived by using the void ratio and the grain size distribution based on the assumptions that the pore size distribution can be expressed by the logarithmic normal distribution as well as the grain size distribution, and that both of the coefficients of variation for the pore size and grain size distributions are same. But the pore size distribution estimated by the void ratio is considered to be the first approximate distribution. In this section, we propose the correction method of pore size distribution when the experimental data is obtained in-situ and/or laboratory water retention tests. Figure 6.11 shows the flow chart for the correction procedure of pore size distribution.

Let's consider the situation where a measured datum, i.e., a set of water content and suction (w, s_u), is obtained by the water retention test, using the simple hanging column apparatus.

Substituting the measured suction s_u into Equation (5.5.6), the maximum diameter d of pipe filled with water corresponding to D_v in the elementary particulate model (see Figure 3.9(c)) is calculated.

On the other hand, substituting calculated d_w and measured water content w into Equation (3.2.19), the corrected distribution parameter λ_v in Equation (3.2.8) can be back-calculated, assuming that the standard deviation ζ_v is not changed and is the same as ζ_s. Then the corrected cumulative percentage of pore size $F_v(d_w)$, i.e., a point $(d, F_v(d_w))$ is obtained according to the calculation procedure as shown in Figure 6.11.

When more measured data sets (w, s_u) are obtained, the corresponding data sets of $(d, F_v(d_w))$ are used to obtain more accurate pore distribution, applying the regression analysis method.

6.4.2 Parallel translation index $I_{\text{pt,s}}$ for soil water characteristic curve

According to the correction procedure shown in Figure 6.11, we can get the more accurate pore size distribution when a measured set or sets of (w, s_u) are obtained. To examine the relation between the grain size distribution, the original pore size distribution and the corrected pore size distribution, the new physical quantity $I_{\text{pt,s}}$ (called the parallel translation index) is introduced for the soil water characteristic curve (Sako and Kitamura, 2006).

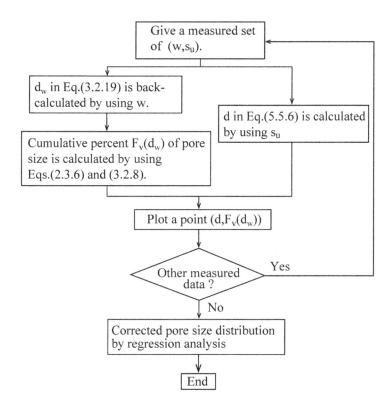

Figure 6.11 Calculation procedure for correction of pore size distribution.

Figure 6.12(a) shows three logarithmic normal distributions, i.e., the grain size distribution, the original pore size distribution and the corrected pore size distribution. The plots of grain size distribution in Figure 6.12(a) are obtained from Table 6.1. The results of the grain size analysis are approximated by the logarithmic normal distribution curve as shown in Figure 6.12(a), applying the regression analysis for the data shown in Table 6.1.

The original pore size distribution is back-calculated by giving the void ratio and using Equation (3.2.18). The volcanic soil called "Shirasu" in Japanese is used as a coarse-grained soil to check the validity of the correction method. Shirasu is geologically defined as the non-welded part of pyroclastic flow deposits. Table 6.2 shows the input parameters used in the back calculation. Table 6.3 shows the calculated distribution parameters of grain size distribution and original pore size distribution, and the characteristic length D_{cha}.

The corrected pore size distribution is obtained by the procedure shown in Figure 6.11. Figure 6.12(b) shows the original and corrected pore size distributions, using five sets of (w, s_u) measured. Firstly, the cumulative

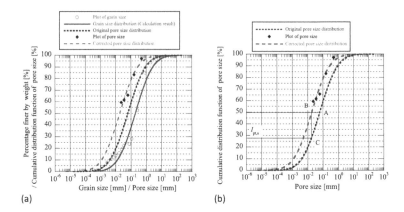

Figure 6.12 Grain size distribution and pore size distributions for soil water characteristic curve (a) Grain size distribution, original pore size distribution and corrected pore size distributin, (b) Parallel translate index $I_{pt,w}$.

Table 6.1 Test Data of Shirasu Obtained from Grain Size Analysis

No.	Parcentage finer by weight [%]	Grain size [mm]
1	4	0.0012
2	7.5	0.0027
3	10.6	0.0052
4	12.1	0.0071
5	13.7	0.0099
6	15.9	0.0162
7	19	0.0242
8	22.1	0.0321
9	23.4	0.075
10	27.9	0.106
11	52.5	0.25
12	68.4	0.425
13	84.8	0.85
14	93.6	2
15	96	4.75
16	98.1	9.5
17	99.1	19

percentage 50% of the original pore size distribution is plotted as point A, and then the cumulative percentage 50% of the corrected pore size distribution is plotted as point B. Finally, point C is plotted on the original pore size distribution as point C. The cumulative percentage at the point C is defined as the parallel translation index, $I_{pt,s}$ for the soil water characteristic curve.

Table 6.2 Values of Input Parameters

Sample		Shirasu
Density of soil particles	[Mg/m³]	2.45
Surface tension of water (at 15°C)	[N/m]	73.48×10^{-3}
Viscosity coefficient of waer (at 15°C)	[Pa·s]	1.138×10^{-3}
The lowest height of probability density function of pipe	[−]	0.159
Void rasio	[−]	1.46

Table 6.3 Values of Parameters for Original Pore Size Distribution and D_{cha}

Mean value of $\ln D_s$	λ_s	−3.997
Standard deviation of $\ln D_s$	ζ_s	2.046
Mean value of D_s [m]	μ_s	1.49×10^{-3}
Standard deviation of D_s [m]	σ_s	1.20×10^{-2}
Mean value of $\ln D_v$	λ_v	−5.128
Standard deviation of $\ln D_v$	ζ_v	2.046
Mean value of D_v [m]	μ_v	4.81×10^{-4}
Standard deviation of D_v [m]	σ_v	3.87×10^{-3}
Coefficient of variation	κ	8.051
Size of D_{cha} [m]	D_{cha}	4.13×10^{-5}
Parcentage finer by weight of D_{cha} [%]	−	23.3

Table 6.4 shows five values of $I_{pt,s}$ for the measured sets of (w, s_u) and their average.

It is found from the calculated results shown in Table 6.4 that the characteristic length D_{cha} corresponds to the diameter of 23.3% finer by weight on the cumulative grain size distribution curve for the soil water characteristic curve of Shirasu. It was found that the linear relation between the coefficient of uniformity U_c and the parallel translation index $I_{pt,s}$ (Sako and Kitamura, 2006).

Table 6.4 Parallel Translation Indices $I_{pt,s}$ for 5 Test Data and Their Average

	$I_{pt,s}$ for SWCC [%]
	23.7
	29.0
	35.0
	31.0
	20.0
Ave.	27.7

REFERENCES

Editorial Committee for Japanese Geotechnical Society Standards (2009). Laboratory
    Testing Standards of Geomaterials, 162 (in Japanese).
Sako, K. and Kitamura, R. (2006). A practical numerical model for seepage behavior
    of unsaturated soil. *Soils and Foundations*, 46(5), 595–604.

Chapter 7

Modeling of pore water and pore air flows by elementary particulate model (EPM)

In this chapter, we will discuss the flow of water and air through unsaturated coarse-grained soil by means of the elementary particulate model (EPM) for the pore structure shown in Figure 3.9(c).

7.1 PERMEABILITY OF FLUID PHASES THROUGH COARSE-GRAINED SOIL

7.1.1 Coefficient of water permeability

The water flow through the narrow tube is regarded as the laminar flow and Poiseuille's law is applied to derive the coefficient of water permeability. Poiseuille's law for the velocity of water flow is expressed by the following equation:

$$v = \frac{\rho_w g}{4\mu_w} \cdot \left(R^2 - r^2 \right) \cdot i \tag{7.1.1}$$

where

v: velocity of laminar water flow,
ρ_w: density of water,
μ_w: viscosity coefficient of water,
R: radius of narrow tube,
i: hydraulic gradient and
g: acceleration of gravity.

The mass flow rate Q_p is obtained as follows, referring to Figure 7.1:

$$Q_p = \int_0^R v \cdot 2\pi r \cdot dr \tag{7.1.2}$$

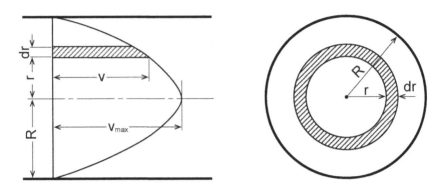

Figure 7.1 Velocity of laminar flow in the narrow tube.

Substituting Equation (7.1.1) into Equation (7.1.2), the following equation is obtained:

$$Q_p = \frac{\rho_w \cdot g}{8\mu_w} \cdot \pi R^4 \cdot i \tag{7.1.3}$$

Then the average velocity is obtained as follows:

$$v_{ave} = \frac{Q}{\pi R^2} = \frac{\rho_w \cdot g}{8\mu_w} \cdot \pi R^2 \cdot i \tag{7.1.4}$$

Equation (7.1.4) is rewritten, using D instead of R.

$$v_{ave} = \frac{\rho_w \cdot g}{8\mu_w} \cdot \left(\frac{D}{2}\right)^2 \cdot i \tag{7.1.5}$$

where D: diameter of narrow tube, i.e., $R = D/2$.

Figure 7.2(a) and (b) show the vertical water flow in the real soil and the elementary particulate model shown in Figure 3.9(c) respectively, where

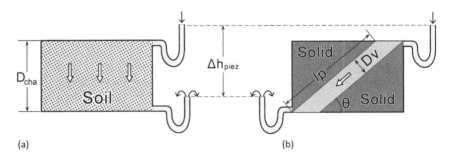

(a)   (b)

Figure 7.2 Water flow in soil and elementary particulate model (EPM) (a) Apparent flow by Darcy's law, (b) Flow in pipe by Poiseulle's law.

the volume of real soil and elementary particulate model are same. When the difference of piezometric head Δh_{piez} is given to the top and bottom plane, the hydraulic gradients are obtained as follows:

$$i_{\text{app}} = \frac{\Delta h_{\text{piez}}}{D_{\text{cha}}} \tag{7.1.6}$$

where i_{app}: apparent hydraulic gradient.

$$i_{\text{EPM}} = \frac{\Delta h_{\text{piez}}}{l_p} = \frac{\Delta h_{\text{piez}}}{D_{\text{cha}}} \sin \theta \tag{7.1.7}$$

where i_{EPM}: hydraulic gradient of pipe in EPM.

When Darcy's law is applied to the real soil shown in Figure 7.2(a), the velocity is obtained as follows:

$$v_{\text{app}} = k_{w,\text{app}} \cdot i_{\text{app}} \tag{7.1.8}$$

where

v_{app}:   apparent velocity of water in real soil,
$k_{w,\text{app}}$:   apparent coefficient of water permeability in real soil and
i_{app}:   apparent hydraulic gradient in real soil obtained by Equation (7.1.6).

The apparent cross-section area A_{app} is obtained as follows, referring to Figure 3.11.

$$A_{\text{app}} = D_v \cdot \left(\frac{D_{\text{cha}}}{\tan \theta} + \frac{D_v}{\sin \theta} \right) \tag{7.1.9}$$

The flow rate Q_{app} calculated by Darcy's law is then obtained as follows, using Equations (7.1.6), (7.1.8) and (7.1.9):

$$Q_{\text{app}} = A_{\text{app}} \cdot v_{\text{app}} = D_v \cdot \left(\frac{D_{\text{cha}}}{\tan \theta} + \frac{D_v}{\sin \theta} \right) \cdot k_{w,\text{app}} \cdot \frac{\Delta h_{\text{piez}}}{D_{\text{cha}}} \tag{7.1.10}$$

where A_{app}: apparent cross-sectional area of real soil.

On the other hand, when Poiseuille's law is applied to the pipe in EPM shown in Figure 7.2(b), the velocity is obtained as follows:

$$v_{\text{EPM}} = \frac{\rho_w g}{8 \mu_w} \cdot \left(\frac{D_v}{2} \right)^2 i_{\text{EPM}} \tag{7.1.11}$$

where

v_{EPM}:   velocity of water in EPM,
$k_{w,\text{EPM}}$:   apparent coefficient of water permeability in EPM and
i_{EPM}:   hydraulic gradient in EPM shown in Equation (7.1.7).

The flow rate Q_{EPM} is then obtained as follows, using Equations (7.1.5), (7.1.7) and (7.1.10):

$$Q_{EPM} = A_{pipe} \cdot V_{EPM} = \pi \cdot \left(\frac{D_v}{2}\right)^2 \cdot \frac{\rho_w g}{8\mu_w} \cdot \left(\frac{D_v}{2}\right)^2 \cdot \frac{\Delta h_{piez}}{D_{cha}} \cdot \sin\theta \qquad (7.1.12)$$

where A_{pipe}: cross-sectional area of pipe in EPM, i.e., $A_{pipe} = \pi \cdot (D_v/2)^2$.

Assuming $Q_{app} = Q_{EPM}$, the following equation is obtained from Equations (7.1.10) and (7.1.12):

$$D_v \cdot \left(\frac{D_{cha}}{\tan\theta} + \frac{D_v}{\sin\theta}\right) \cdot k_{w,app} \cdot \frac{\Delta h_{piez}}{D_{cha}}$$

$$= \pi \cdot \left(\frac{D_v}{2}\right)^2 \cdot \frac{\rho_w g}{8\mu_w} \cdot \left(\frac{D_v}{2}\right)^2 \cdot \frac{\Delta h_{piez}}{D_{cha}} \cdot \sin\theta \qquad (7.1.13)$$

Equation (7.1.13) is rewritten as follows:

$$k_{w,app} = \frac{\rho_w g \cdot D_v^3 \cdot \pi \cdot \sin\theta}{128\mu_w \left(\dfrac{D_{cha}}{\tan\theta} + \dfrac{D_v}{\sin\theta}\right)} = \varphi_{app}(D_v, \theta) \qquad (7.1.14)$$

where $\varphi_{app}(D_v, \theta)$: function of D_v and θ.

As the coefficient of water permeability k_w is considered to be the average of $k_{w,app}$, the following equation is obtained, referring to Equation (2.4.12):

$$k_w = E[k_{w,app}] = \int_0^{d_w} \int_{-\pi/2}^{\pi/2} \varphi_{app}(D_v, \theta) \cdot f(D_v, \theta) d\theta dD_v \qquad (7.1.15)$$

As D_v and θ are independent of each other, Equation (7.1.15) is rewritten as follows, referring to Equation (2.4.13):

$$k_w = \int_0^{d_w} \int_{-\pi/2}^{\pi/2} \varphi_{app}(D_v, \theta) \cdot f_v(D_v) \cdot f_c(\theta) d\theta dD_v \qquad (7.1.16)$$

where
  $f_v(D_v)$: pore size distribution function (seeing Equation (3.2.8)) and
  $f_c(\theta)$: probability density function of predominant flow direction in EMP (seeing Equation (3.2.10)).

Substituting Equation (7.1.14) into Equation (7.1.16), the following equation is obtained:

$$k_w = \int_0^{d_w} \int_{-\pi/2}^{\pi/2} \frac{\rho_w g \cdot D_v^3 \cdot \pi \cdot \sin\theta}{128\mu_w \left(\dfrac{D_{cha}}{\tan\theta} + \dfrac{D_v}{\sin\theta}\right)} \cdot f_v(D_v) \cdot f_c(\theta) d\theta dD_v \qquad (7.1.17)$$

The intrinsic permeability k_w with respect to water in soil is obtained as follows:

$$K_w = \frac{\mu_w}{\rho_w \cdot g} k_w = \int_0^{d_w} \int_{-\pi/2}^{\pi/2} \frac{D_v^3 \cdot \pi \cdot \sin\theta}{128 \left(\frac{D_{cha}}{\tan\theta} + \frac{D_v}{\sin\theta} \right)} f_v(D_v) f_c(\theta) d\theta dD_v \quad (7.1.18)$$

As d_w and D_{cha}, and the probability density functions $f_v(D_v)$ and $f_c(\theta)$ are known, the coefficient of water permeability and/or intrinsic permeability can be numerically calculated by Equations (7.1.17) and (7.1.18). Figure 7.3 shows the calculation procedure to obtain the relation between the water content and the coefficient of water permeability for unsaturated–saturated soil, which is similar to that of soil water characteristic curve shown in Figure 6.2.

The coefficient of water permeability obtained by Equation (7.1.17) is implicitly considered to be the coefficient for the vertical direction, referring to Figure 7.2. It is well known that the coefficient of water permeability

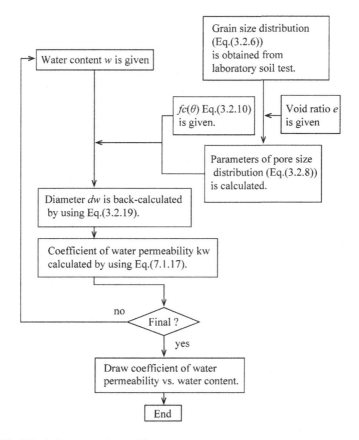

Figure 7.3 Calculation procedure of k_w.

is generally anisotropic. The anisotropy of water permeability can be estimated by shifting the distribution of predominant flow direction defined by Equation (3.2.10) in the elementary particulate model (EPM). For example, the distribution of predominant flow direction shown in Figure 7.4 is used for that of horizontal direction instead of Figure 3.10.

7.1.2 Coefficient of air permeability

Assuming that Darcy's law and Poiseuille's law are available for the air phase in soil, the following equations can be obtained, referring to Equation (7.1.17). As the air is filled with the pipe larger than d_w, Equation (7.1.16) is modified to apply the coefficient of air permeability as follows:

$$k_a = \int_{d_w}^{\infty} \int_{-\pi/2}^{\pi/2} \frac{\rho_a g \cdot D_v^3 \cdot \pi \cdot \sin\theta}{128\mu_a \left(\dfrac{D_{cha}}{\tan\theta} + \dfrac{D_v}{\sin\theta}\right)} \cdot f_v(D_v) \cdot f_c(\theta) d\theta dd D_v \qquad (7.1.19)$$

where

   ρ_a: density of air and
   μ_a: viscosity coefficient of air.

7.2 CORRECTION OF PORE SIZE DISTRIBUTION FOR COEFFICIENT OF WATER PERMEABILITY

7.2.1 Correction method for coefficient of water permeability

The pore size distribution for the permeability estimated by using the void ratio, Equation (3.2.8), is considered to be the first approximate distribution

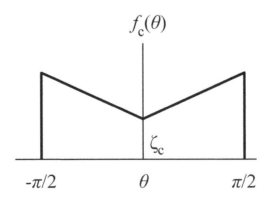

Figure 7.4 Anisotropy of coefficient of water permeability.

for the permeability. We can then correct the pore size distribution for the permeability when the experimental data is obtained in-situ and/or laboratory water permeability tests as well as in the case of the soil water characteristic curve discussed in Section 6.4. Figure 7.5 shows the flow chart of the correction procedure of pore size distribution for permeability (Sako and Kitamura, 2006).

Let's consider the situation where an experimental datum, i.e., a pair of water content and coefficient of water permeability (w, k_w), is obtained by the water permeability test, for example, the constant head permeability test.

Substituting the suction k_w into Equation (7.1.17), the maximum diameter d_w of the pipe filled with water corresponding to D_v in the elementary particulate model (see Figure 3.9(c)) is back-calculated.

On the other hand, substituting d_w and water content w into Equation (3.2.19), the corrected distribution parameter λ_v in Equation (3.2.8) can be back-calculated, assuming that the standard deviation ζ_v is not changed and is the same as ζ_s. Then the corrected cumulative percent of pore size $F_v(d_w)$, i.e., a point $(d, F_v(d_w))$ is obtained as shown in Figure 7.5.

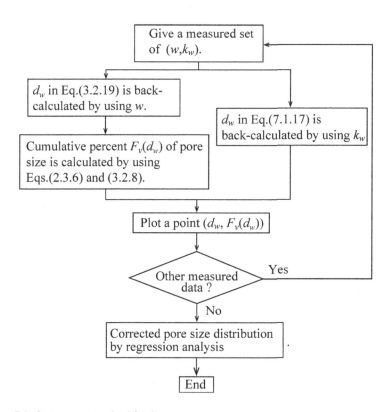

Figure 7.5 Correction method for k_w.

When more experimental data (w,k_w) are obtained, the corresponding dataset of $(d, F_v(d_w))$ are used to obtain more accurate pore distribution, applying the regression analysis method.

7.2.2 Parallel translation index $I_{pt,w}$ for coefficient of water permeability

According to the correction procedure shown in Figure 7.5, we can get a more accurate pore size distribution for the coefficient of water permeability as well as the soil water characteristic curve when a measured set or sets of (w,k_w) are obtained, where w and k_w are respectively the water content and the coefficient of water permeability. To examine the relation between the grain size distribution, the original pore size distribution and the corrected pore size distribution for the coefficient of water permeability, a new physical quantity $I_{pt,w}$ called the parallel translation index is introduced for the coefficient of water permeability (Sako and Kitamura, 2006).

Figure 7.6(a) shows three logarithmic normal distributions, i.e., the grain size distribution, the original pore size distribution and the corrected pore size distribution. The plots of grain size distribution in Figure 7.6(a) are obtained from Table 6.1 as well as Figure 6.12. The results of grain size analysis are approximated by the logarithmic normal distribution curve as shown in Figure 7.6(a), applying the regression analysis for the data shown in Table 6.1.

The original pore size distribution is back-calculated by giving the void ratio and using Equation (3.2.18), as well as Figure 6.12. The back-calculation is carried out by using the values shown in Table 6.1. The calculated distribution parameters of grain size distribution and original pore size distribution and the characteristic length D_{cha} are same as those shown in Table 6.3.

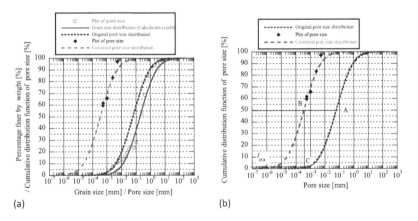

Figure 7.6 Grain size distribution and pore size distributions for coefficient of water permeability (a) Grain size distribution, original pore size distribution and corrected pore size distribution, (b) Parallel translate index $I_{pt,s}$

Table 7.1 Parallel Translation Indices $I_{pt,w}$ for
Five Test Datasets and Their Average

$I_{pt,w}$ for k [%]
0.6
0.6
1
0.7
0.2
Ave.   0.6

The corrected pore size distribution is obtained by the procedure shown in Figure 7.5. Figure 7.6(b) shows the original and corrected pore size distributions, using five sets of (w,k_w) measured. The cumulative percentage at the point C is defined as the parallel translation index, $I_{pt,w}$ for the coefficient of water permeability. Table 7.1 shows 5 values of $I_{pt,w}$ for the measured sets of (w,k_w) and their average. It is found from the calculated results shown in Table 7.1 that the characteristic length D_{cha} corresponds to the diameter of 0.6% finer by weight on the cumulative grain size distribution curve for the coefficient of water permeability of Shirasu. Comparing the values in Table 7.1 with those in Table 6.4, it might be considered that the pore water contributed to the water flow is much less than the water retention through the soil block. It was found that there was a linear relation between the coefficient of uniformity and $I_{pt,w}$ (Sako and Kitamura, 2006).

7.3 GOVERNING EQUATION FOR SATURATED–UNSATURATED SEEPAGE FLOW IN SOIL

7.3.1 Derivation of governing equation

Figure 7.6 shows a minute element which contains a few soil particles. It is assumed that a soil particle is rigid and that particle crushing does not occur. The x- and y-axes are horizontal and the z-axis is vertical in the reference coordinate. The lengths of minute element are Δx, Δy, Δz in the direction of x-, y- and z-axes. Darcy's law can be then expressed as follows:

$$v_x = -k_{wx}\left(e, h_{pre}\right)\frac{\partial h_{piez}}{\partial x} \tag{7.3.1}$$

$$v_y = -k_{wy}\left(e, h_{pre}\right)\frac{\partial h_{piez}}{\partial y} \tag{7.3.2}$$

$$v_z = -k_{wz}\left(e, h_{pre}\right)\frac{\partial h_{piez}}{\partial z} \tag{7.3.3}$$

where

v_x: apparent velocity of pore water in the direction of x-axis,
v_y: apparent velocity of pore water in the direction of y-axis,
v_z: apparent velocity of pore water in the direction of z-axis,
e: void ratio,
h_{pre}:   pressure head,
$k_{wx}(e, h_{pre})$: coefficient of water permeability in the direction of x-axis, which is expressed as a function of e and h_{pre},
$k_{wy}(e, h_{pre})$: coefficient of water permeability in the direction of y-axis, which is expressed as a function of e and h_{pre},
$k_{wz}(e, h_{pre})$: coefficient of water permeability in the direction of z-axis, which is expressed as a function of e and h_{pre},
h_{piez}:   piezometric head.

The mass of flow into and flow out per unit time m_{in} and m_{out} in the minute element shown in Figure 7.7 are obtained as follows:

$$m_{in} = \rho_w v_x \Delta y \Delta z + \rho_w v_y \Delta z \Delta x + \rho_w v_z \Delta x \Delta y \tag{7.3.4}$$

$$m_{out} = \left(\rho_w v_x + \frac{\partial(\rho_w v_x)}{\partial x} \Delta x \right) \Delta y \Delta z + \left(\rho_w v_y + \frac{\partial(\rho_w v_y)}{\partial y} \Delta y \right) \Delta z \Delta x$$
$$+ \left(\rho_w v_z + \frac{\partial(\rho_w v_z)}{\partial z} \Delta z \right) \Delta x \Delta y \tag{7.3.5}$$

Then the net mass remained in the minute element Δm_w is obtained as follows:

$$\Delta m_w = m_{in} - m_{out} = -\left(\frac{\partial(\rho_w v_x)}{\partial x} + \frac{\partial(\rho_w v_y)}{\partial y} + \frac{\partial(\rho_w v_z)}{\partial z} \right) \Delta x \Delta y \Delta z$$
$$= -\left(\frac{\partial(\rho_w v_x)}{\partial x} + \frac{\partial(\rho_w v_y)}{\partial y} + \frac{\partial(\rho_w v_z)}{\partial z} \right) \Delta V \tag{7.3.6}$$

where ΔV: volume of minute element shown in Figure 7.7.

On the other hand, as the net mass Δm_w is the change of pore water mass per unit time in the minute element, Δm_w can be expressed as follows:

$$\Delta m_w = \frac{\partial(\rho_w \Delta V_w)}{\partial t} \tag{7.3.7}$$

where ΔV_w: volume of pore water in the minute element shown in Figure 7.7.
Referring to Equations (3.1.8) and Figure 3.3, the following equation is obtained.

$$\Delta V_w = n_w \cdot \Delta V \tag{7.3.8}$$

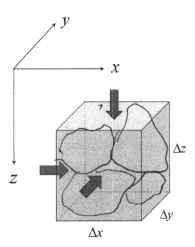

Figure 7.7 Flow of pore water through minute element in soil.

where n_w: volumetric water content.

Substituting Equation (7.3.8) into Equation (7.3.7), the following equation is obtained:

$$\Delta m_v = \frac{\partial(\rho_w \cdot n_w \cdot \Delta V)}{\partial t} = \frac{\partial(\rho_w \cdot n_w)}{\partial t} \cdot \Delta V \tag{7.3.9}$$

Assuming that there are no sources in the minute element, the following equation is obtained by equating Equations (7.3.6) and (7.3.9):

$$\frac{\partial(\rho_w \cdot n_w)}{\partial t} = -\left(\frac{\partial(\rho_w v_x)}{\partial x} + \frac{\partial(\rho_w v_y)}{\partial y} + \frac{\partial(\rho_w v_z)}{\partial z}\right) \tag{7.3.10}$$

Assuming that the pore water is incompressible and that the density of pore water is constant, Equation (7.3.10) is rewritten as follows:

$$\frac{\partial n_w}{\partial t} = -\left(\frac{\partial v_x}{\partial x} + \frac{\partial v_y}{\partial y} + \frac{\partial v_z}{\partial z}\right) \tag{7.3.11}$$

Substituting Equations (7.3.1)–(7.3.3) into Equation (7.3.11), the following equation is obtained:

$$\frac{\partial n_w}{\partial t} = \frac{\partial\left(k_{wx}(e, h_{\text{pre}})\frac{\partial h_{\text{piez}}}{\partial x}\right)}{\partial x} + \frac{\partial\left(k_{wy}(e, h_{\text{pre}})\frac{\partial h_{\text{piez}}}{\partial y}\right)}{\partial y}$$
$$+ \frac{\partial\left(k_{wz}(e, h_{\text{pre}})\frac{\partial h_{\text{piez}}}{\partial z}\right)}{\partial z} \tag{7.3.12}$$

The relations between the piezometric head h_{piez} and the pressure head h_{pcw} in the x-, y- and z- direction are expressed as follows:

For the x- and y- direction (horizontal direction),

$$h_{piez} = h_{pre} \tag{7.3.13}$$

For the z- direction (vertical direction),

$$h_{piez} = h_{pre} + z \tag{7.3.14}$$

Substituting Equations (7.3.13) and (7.3.14) into Equation (7.3.12), the following equation is obtained.

$$\frac{\partial n_w}{\partial t} = \frac{\partial\left(k_{wx}\left(e, h_{pre}\right)\frac{\partial h_{pre}}{\partial x}\right)}{\partial x} + \frac{\partial\left(k_{wy}\left(e, h_{pre}\right)\frac{\partial h_{pre}}{\partial y}\right)}{\partial y}$$

$$+ \frac{\partial\left(k_{wz}\left(e, h_{pre}\right)\frac{\partial h_{pre}}{\partial z} + k_{wz}\left(e, h_{pre}\right)\right)}{\partial z} \tag{7.3.15}$$

Referring to Equations (3.1.4), (3.1.7) and (3.1.8), the following equation is obtained:

$$n_w = n \cdot S_r \tag{7.3.16}$$

where n_w: water porosity (volumetric water content), n: porosity, S_r: degree of saturation.

Applying the chain rule to Equation (7.3.16), the following equation is obtained:

$$\frac{\partial n_w}{\partial t} = \frac{\partial n_w}{\partial h_{pre}} \cdot \frac{\partial h_{pre}}{\partial t} = \frac{\partial\left(n \cdot S_r\right)}{\partial h_{pre}} \cdot \frac{\partial h_{pre}}{\partial t}$$

$$= \left(n \cdot \frac{\partial S_r}{\partial h_{pre}} + S_r \cdot \frac{\partial n}{\partial h_{pre}}\right) \cdot \frac{\partial h_{pre}}{\partial t} \tag{7.3.17}$$

where $\partial n_w / \partial h_{pre}$: specific moisture coefficient obtained from the soil water characteristic curve.

Substituting Equation (7.3.17) into Equation (7.3.15), the following partial differential equation is obtained:

$$\left(n \cdot \frac{\partial S_r}{\partial h_{\mathrm{pre}}} + S_r \cdot \frac{\partial n}{\partial h_{\mathrm{pre}}}\right) \cdot \frac{\partial h_{\mathrm{pre}}}{\partial t} = \frac{\partial \left(k_{wx}\left(e, h_{\mathrm{pre}}\right)\dfrac{\partial h_{\mathrm{pre}}}{\partial x}\right)}{\partial x}$$

$$+ \frac{\partial \left(k_{wy}\left(e, h_{\mathrm{pre}}\right)\dfrac{\partial h_{\mathrm{pre}}}{\partial y}\right)}{\partial y} \qquad (7.3.18)$$

$$+ \frac{\partial \left(k_{wz}\left(e, h_{\mathrm{pre}}\right)\dfrac{\partial h_{\mathrm{pre}}}{\partial z} + k_{wz}\left(e, h_{\mathrm{pre}}\right)\right)}{\partial z}$$

where

$\partial S_r/\partial h_{\mathrm{pre}}$: specific saturation coefficient obtained from the soil water characteristic curve, i.e, the gradient of suction–degree of saturation curve discussed in Section 6.2,

$\partial n/\partial h_{\mathrm{pre}}$: specific storage coefficient obtained from the gradient of the curve projected on the $e{\sim}p$ plane in Figure 3.7 for unsaturated soil and the coefficient of volume change for saturated soil.

The suction–degree of saturation curve used to obtain $\partial S_r/\partial h_{\mathrm{pre}}$ corresponds to the curve projected on the $w{\sim}$negative pressure plane in Figure 3.7. The specific storage coefficients $\partial n/\partial h_{\mathrm{pre}}$ under unsaturated conditions and saturated conditions respectively correspond to the gradient of the curve projected on the $e{\sim}$negative pressure plane and compression curve in Figure 3.7.

Equation (7.3.18) is the general basic equation of seepage of pore water in saturated–unsaturated soil. When the specific saturation coefficient and the specific storage coefficient, which are the input parameters of the basic equation for the saturated–saturated seepage, are obtained from the in-situ and/or laboratory soil tests, $h_{\mathrm{pre}}(x,y,z;t)$ is obtained, solving Equation (7.3.18) with initial and boundary conditions, i.e., the distribution of pressure head with the change in time in the 3-D space is known.

7.3.2 Permeability function

As the permeability function $k_w(e, h_{\mathrm{pre}})$ in Equation (7.3.18) for saturated soil depends only on the void ratio, $k_w(e, h_{\mathrm{pre}})$ is rewritten as follows:

For saturated soil,

$$k_{w,\mathrm{sat}}\left(e, h_{\mathrm{pre}}\right) = k_w\left(e\right) \qquad (7.3.19)$$

The permeability function $k_w(e, h_{pre})$ in Equation (7.3.18) for unsaturated soil can be transformed to $k_w(e, w)$ because the negative pressure head h_{pre} is related to the water content w through the soil water characteristic curve, i.e., the permeability functions with respect to the pressure head and the water content are compatible with each other as follows.

For unsaturated soil,

$$k_{w,unsat}\left(e, h_{pre}\right) \Leftrightarrow k_{w,unsat}\left(e, w\right) \qquad (7.3.20)$$

Figure 7.8 shows the schematic relation between the void ratio, the water content and the coefficient of water permeability in the $e \sim w \sim k_w$ space, where the maximum and minimum void ratios are shown on the e-axis as well as those in Figure 3.4. The line A–B on $e = w \cdot G_s$ plane is the relation between k_w and void ratio (or water content) for the saturated soil. The line A'–B' is the projected line of the line A–B on $w \sim k_w$ plane. The line A''–B'' is the projected line of the line A–B on $e \sim k_w$ plane. A point C on A–B line denotes an arbitrary coefficient of saturated water permeability and the curve C–D shows the permeability function in $e \sim w \sim k_w$ space for unsaturated condition. The curve C'–D' is the projected curve on $w \sim k_w$ plane. The curve C''–D'' is the projected curve on $e \sim k_w$ plane. As the void ratio at the point C in Figure 7.8 changes with the change in suction (= water content), the void ratio at the point D is not same as that at the point C.

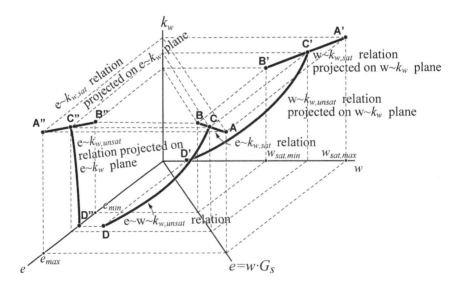

Figure 7.8 $e \sim w \sim k_w$ relation in 3-D space.

7.3.3 Governing equation under limiting conditions

Assuming the volume change can be neglected for coarse soil, the following equation is obtained:

$$\frac{\partial n}{\partial h_{\mathrm{pre}}} = 0 \qquad (7.3.21)$$

Substituting Equation (7.3.21) into Equation (7.3.18), the following equation is obtained:

$$n \cdot \frac{\partial S_r}{\partial h_{\mathrm{pre}}} \cdot \frac{\partial h_{\mathrm{pre}}}{\partial t} = \frac{\partial \left(k_{wx}\left(e, h_{\mathrm{pre}}\right) \frac{\partial h_{\mathrm{pre}}}{\partial x} \right)}{\partial x} + \frac{\partial \left(k_{wy}\left(e, h_{\mathrm{pre}}\right) \frac{\partial h_{\mathrm{pre}}}{\partial y} \right)}{\partial y}$$

$$+ \frac{\partial \left(k_{wz}\left(e, h_{\mathrm{pre}}\right) \frac{\partial h_{\mathrm{pre}}}{\partial z} + k_{wz}\left(e, h_{\mathrm{pre}}\right) \right)}{\partial z} \qquad (7.3.22)$$

As the degree of saturation S_r is 1, the volumetric water content n_w is equal to the porosity n, and the coefficient of water permeability is the function of the void ratio for saturated soil, Equation (7.3.22) is rewritten as follows:

$$\frac{\partial n}{\partial h_{\mathrm{pre}}} \cdot \frac{\partial h_{\mathrm{pre}}}{\partial t} = k_{wx,\mathrm{sat}}\left(e\right) \cdot \frac{\partial^2 h_{\mathrm{pre}}}{\partial x^2} + k_{wy,\mathrm{sat}}\left(e\right) \cdot \frac{\partial^2 h_{\mathrm{pre}}}{\partial y^2}$$

$$+ k_{wz,\mathrm{sat}}\left(e\right) \cdot \frac{\partial^2 h_{\mathrm{pre}}}{\partial z^2} \qquad (7.3.23)$$

When the coefficient of water permeability is isotropic, Equation (7.3.23) is rewritten as follows:

$$\frac{\partial n}{\partial h_{\mathrm{pre}}} \cdot \frac{\partial h_{\mathrm{pre}}}{\partial t} = k_{w,\mathrm{sat}}\left(e\right) \cdot \left(\frac{\partial^2 h_{\mathrm{pre}}}{\partial x^2} + \frac{\partial^2 h_{\mathrm{pre}}}{\partial y^2} + \frac{\partial^2 h_{\mathrm{pre}}}{\partial z^2} \right) \qquad (7.3.24)$$

In the case of one-dimensional seepage in saturated soil, Equation (7.3.24) is rewritten as follows:

$$\frac{\partial n}{\partial h_{\mathrm{pre}}} \cdot \frac{\partial h_{\mathrm{pre}}}{\partial t} = k_{w,\mathrm{sat}}\left(e\right) \cdot \frac{\partial^2 h_{\mathrm{pre}}}{\partial z^2} \qquad (7.3.25)$$

Now let's introduce the following equation which is often used in the conventional soil mechanics:

$$m_v = -\frac{\Delta n}{\Delta p} \qquad (7.3.26)$$

where

m_v: coefficient of volume compressibility,

Δn: change in porosity and

Δp: change in effective pressure.

Applying the effective stress equation for saturated soil derived by Terzaghi and Peck (1948), the following equation is obtained:

$$\Delta p = -\Delta u = -\Delta\left(\rho_w g h_{pre}\right) \tag{7.3.27}$$

where Δu: change in pore water pressure.

Substituting Equation (7.3.27) into Equation (7.3.26), the following equation is obtained:

$$m_v = \frac{\Delta n}{\Delta\left(\rho_w g h_{pre}\right)} \tag{7.3.28}$$

Equation (7.3.28) is rewritten as follows:

$$\Delta n = \rho_w g \cdot m_v \cdot \Delta h_{pre} \tag{7.3.29}$$

The following equation is then obtained:

$$\frac{\Delta n}{\Delta h_{pre}} = \rho_w g \cdot m_v \tag{7.3.30}$$

Substituting Equation (7.3.30) into Equation (7.3.25), the following equation is obtained:

$$\rho_w g \cdot m_v \cdot \frac{\partial h_{pre}}{\partial t} = k_{w,sat}(e)\frac{\partial^2 h_{pre}}{\partial z^2} \tag{7.3.31}$$

Equation (7.3.31) is rewritten as follows:

$$\frac{\partial h_{pre}}{\partial t} = \frac{k_{w,sat}(e)}{\rho_w g \cdot m_v}\frac{\partial^2 h_{pre}}{\partial z^2} = c_v\frac{\partial^2 h_{pre}}{\partial z^2} \tag{7.3.32}$$

where

$$c_v = \frac{k_{w,sat}(e)}{\rho_w g \cdot m_v} \tag{7.3.33}$$

c_v: coefficient of consolidation.

Equation (7.3.32) is the popular basic equation of one-dimensional consolidation for saturated soil derived by Terzaghi and Peck (1948). Therefore it is found that the basic equation of one-dimensional consolidation can be

derived from Equation (7.3.18) which is the general basic equation for saturated–unsaturated seepage, and that the basic equation for one-dimensional
consolidation is one example to express the seepage behavior of saturated
isotropic soil coupled with the constitutive equation, Equation (7.3.26).

Furthermore, the following equation is obtained from Equation (7.3.24)
for the steady seepage state, because $\partial h_{\text{piez}}/\partial t = 0$, and the pressure head
can be related to the piezometric head as shown in Equations (7.3.13) and
(7.3.14):

$$\frac{\partial^2 h_{\text{piez}}}{\partial x^2} + \frac{\partial^2 h_{\text{piez}}}{\partial y^2} + \frac{\partial^2 h_{\text{piez}}}{\partial z^2} = 0 \tag{7.3.34}$$

Equation (7.3.34) is the popular equation called "Laplace's equation" and
is used for the flow net analysis in the conventional soil mechanics.

REFERENCES

Sako, K. and Kitamura, R. (2006). A practical numerical model for seepage behavior
of unsaturated soil. *Soils and Foundations*, 46(5), 595–604.
Terzaghi, K., and Peck, R.B. (1948). *Soil Mechanics in Engineering Practice*. John
Wiley & Sons, 233–242.

Chapter 8

Stability analysis by proposed model

8.1 FRICTION LAW

8.1.1 Friction law of a solid body

Figure 8.1 shows a solid body on the horizontal plane, where the vertical force F_N due to the self-weight of the body acts on the horizontal plane. When the horizontal force F_T is gradually increased in the range of less than the resistance force F_R, the body does not move. However, when F_T reaches to F_R, the body begins to move. This condition is called the limit equilibrium condition under which the following equation is satisfied:

$$F_T = F_R = \mu_{stat} \cdot F_N \qquad (8.1.1)$$

where

F_T: tangential force,
F_N: normal force,
F_R: resistance force and
μ_{stat}: coefficient of static friction.

Equation (8.1.1) is rewritten as follows:

$$\mu_{stat} = \frac{F_T}{F_N} = \tan \phi_{stat} \qquad (8.1.2)$$

where ϕ_{stat}: static friction angle.

Both Equations (8.1.1) and (8.1.2) represent the friction law of a body.
Figure 8.2 shows the critical state of slipping where the solid body on the plane just begins to move as the inclination angle of plane is slowly

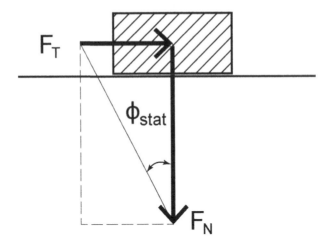

Figure 8.1 Limit equilibrium condition.

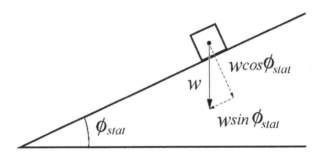

Figure 8.2 Slipping condition of body on the inclination plane.

increased. The force equilibrium of the direction parallel to the inclined plane in this critical state is expressed as follows:

$$W \cdot \sin\varphi_{stat} = \mu_{stat} \cdot W \cdot \cos\varphi_{stat} \tag{8.1.3}$$

where
  W: weight of solid body,
  μ_{stat}: coefficient of static friction and
  φ_{stat}: critical angle of inclination plane.

Equation (8.1.3) is rewritten as follows:

$$\mu_{stat} = \frac{W \cdot \sin\varphi_{stat}}{W \cdot \cos\varphi_{stat}} = \frac{F_T}{F_N} = \tan\varphi_{stat} \tag{8.1.4}$$

Equation (8.1.4) is defined as the slipping condition of a solid body.

8.1.2 Friction law of particulate soil block

A soil particle in a soil block usually has more than two contact points, as shown in Figure 4.6. As the particulate soil block is considered to be a friction material, it is considered that the friction law of a single body shown in Figure 8.2 can be extended to a particulate soil block, where Equation (8.1.4) is transformed to the following equation:

$$\mu_{ave} = \frac{F_T}{F_N} = \tan\varphi_{ave} \tag{8.1.5}$$

where

μ_{ave}: coefficient of average friction in soil block and
φ_{ave}: average friction angle in soil block.

The angle of repose is applied to estimate φ_{ave} in Equation (8.1.5), i.e, the average friction angle at the critical state in soil block is assumed to be same as φ_{ave} as follows:

$$\varphi_{ave} = \varphi_{repose} \tag{8.1.6}$$

where φ_{repose}: angle of repose obtained by the experiment shown in Figure 8.3.
Furthermore, it is assumed that F_T/F_N in Equation (8.1.5) is estimated by $F_{\beta,N}$ and $F_{\beta,T}$ shown in Equations (5.7.1) and (5.7.2) as follows:

$$\frac{F_T}{F_N} = \max\left(\frac{F_{\beta,T}}{F_{\beta,N}}\right) \tag{8.1.7}$$

where

$$F_{\beta,N} = F_{\beta,grav,N} + F_{\beta,seep,N} + F_{\beta,matr,N} + F_{\beta,ext,N} \tag{5.7.1bis}$$

$$F_{\beta,T} = F_{\beta,grav,T} + F_{\beta,seep,T} + F_{\beta,ext,T} \tag{5.7.2bis}$$

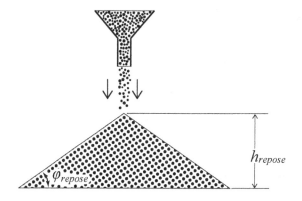

Figure 8.3 Experiment to obtain the angle of repose.

$F_{\beta,T}/F_{\beta,N}$ in Equation (8.1.7) is called the inter-particle stress ratio in the following sections in this chapter.

Consequently, the slipping condition of particulate soil block at the critical state is expressed as follows:

$$\max\left(\frac{F_{\beta,T}}{F_{\beta,N}}\right) = \tan\varphi_{\text{repose}} \qquad (8.1.8)$$

8.2 POTENTIAL SLIP PLANE

Let's consider the conventional triaxial compression test on unsaturated soil under the constant confining pressure. The inter-particle stress state after isotropic compression is expressed as the point $(F_{N,\min}, F_T=0)$ on the $F_N \sim F_T$ plane shown in Figure 8.4. As the axial pressure is increased, the diameter of Mohr's stress circles with respect to the inter-particle stress are increased as shown in the circles C1 and C2, and finally the circle C3 has contact with the failure line.

Figure 8.5 shows the Mohr's stress circle C2 taken from Figure 8.4. A stress point $(F_{\beta,N}, F_{\beta,T})$ with the angle 2β on the circle's perimeter in Figure 8.5 corresponding to the normal and tangential components of inter-particle stress vector with the inclination angle β in the specimen as shown in Figures 5.20 and 5.21. In other words, all of the inter-particle stress vectors in the specimen can be expressed by the points on the circle's perimeter. When the angle of the tangential line for the circle

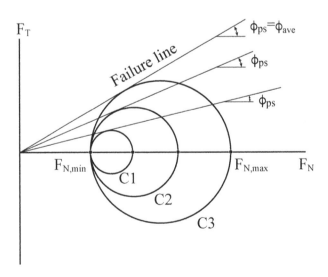

Figure 8.4 Change in Mohr's stress circles during triaxial compression test.

C2 drawn from the origin is denoted by ϕ_{ps}, the following equations are derived:

$$\max\left(\frac{F_{\beta,T}}{F_{\beta,N}}\right) = \tan\phi_{ps} \tag{8.2.1}$$

$$\beta_{ps} = \frac{\pi}{4} + \frac{\phi_{ps}}{2} \tag{8.2.2}$$

where

$F_{\beta,N}$:   normal component of inter-particle stress vector $\vec{F}_\beta$ on the plane with inclination angle β,

$$F_{\beta,N} = F_{\beta,\text{grav},N} + F_{\beta,\text{seep},N} + F_{\beta,\text{matr},N} + F_{\beta,\text{ext},N} \tag{5.7.1bis}$$

$F_{\beta,T}$:   tangential component of inter-particle stress vector $\vec{F}_\beta$ on the plane with inclination angle β,

$$F_{\beta,T} = F_{\beta,\text{grav},T} + F_{\beta,\text{seep},T} + F_{\beta,\text{ext},T} \tag{5.7.2bis}$$

β_{ps}:   angle of plane where inter-particle stress vector ratio $F_{\beta,T}/F_{\beta,N}$ is maximum and

ϕ_{ps}:   angle of tangential line shown in Figure 8.5

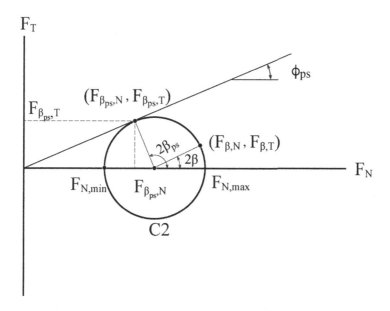

Figure 8.5 Relation between tangential plane with contact angle β and Mohr's stress circle.

In the proposed model, the plane defined by Equation (8.2.1) is called the potential slip plane, and β_{ps} defined by Equation (8.2.2) is called the inclination angle of the potential slip plane for the stress state expressed by the circle C2. The potential slip plane for the circles C1 and C3 can also be obtained as shown in Figure 8.4. φ_{ps} for the circle C3 is the same as the average friction angle φ_{ave}.

Figure 8.6 shows the Mohr's stress circle C3 taken from Figure 8.4. The circle C3 represents the stress state at failure. The following equation is then obtained:

$$\max\left(\frac{F_{\beta,T}}{F_{\beta,N}}\right) = \tan\varphi_{ps,max} = \tan\varphi_{ave} = \mu_{ave} \tag{8.2.3}$$

where ϕ_{ave}: average friction angle in soil block defined by Equations (8.1.5) and (8.1.6), i.e., equal to angle of repose, corresponding to angle of internal friction (angle of shearing resistance) used in the conventional soil mechanics.

8.3 APPARENT COHESION DUE TO SURFACE TENSION

Figure 8.7(a) shows the Mohr–Coulomb failure envelope proposed by Fredlund et al. (2012) for unsaturated soil. Figure 8.7(b) shows the relation

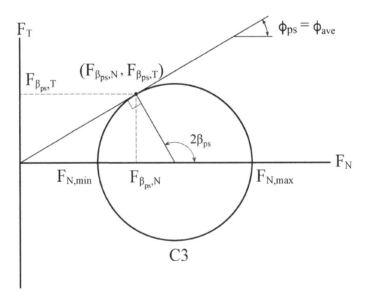

Figure 8.6 Relation between tangential plane with contact angle β and Mohr's stress circle at failure.

between the matric suction and the shear stress when the net normal stress is zero, i.e., the intersection line of failure envelope on the shear stress τ and matric suction $s_u = u_a - u_w$. As the apparent cohesion plays the most important role in analyzing the shear properties in the current soil mechanics, we briefly discuss it in this section, although the apparent cohesion is never related to the proposed model.

It is found from Figure 8.7 that the apparent cohesion c for unsaturated soil is expressed as follows:

$$c = c_0 + c_1 \tag{8.3.1}$$

where
c_0: apparent cohesion of dry or saturated soil and
c_1: apparent cohesion due to surface tension.

Because the microscopic mechanism for the apparent cohesion due to surface tension has not been made clear on the failure envelope, as shown in Figure 8.7(a), it is boldly assumed that c_1 in Equation (8.3.1) corresponds to F_T in Equation (8.1.5). The following equation is then obtained:

$$c_1 = \mu_{ave} \cdot F_N \tag{8.3.2}$$

F_N in Equation (8.3.2) is considered to correspond to the inter-particle stress vector due to surface tension and then the following equation is obtained, referring to Equations (4.1.9), (4.2.13), (4.3.7) and (5.5.20):

$$F_N = F_{matr,N} = N_{ca} \cdot \left(2\pi \cdot T_s \cdot E[r_i'] + \pi \cdot s_u \cdot E[r_i'^2] \right) \tag{8.3.3}$$

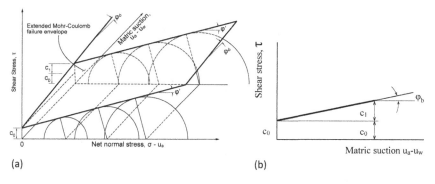

Figure 8.7 Mohr–Coulomb failure envelope for unsaturated soils (a) Failure envelop in 3-D space, (b) Failure line on $\tau \sim (u_a - u_w)$ plane.

(After Fredlund et al., 2012, modified).

Substituting Equation (8.3.3) into Equation (8.3.2), the following equation is obtained:

$$c_1 = \mu_{ave} \cdot N_{ca} \cdot \left(2\pi \cdot T_s \cdot E[r_i'] + \pi \cdot s_u \cdot E\left[r_i'^2\right] \right) \tag{8.3.4}$$

Equation (8.3.4) is the apparent cohesion due to surface tension derived from the proposed model and a useful equation to carry out the slope stability analysis with the rainfall, although Equation (8.3.4) is never used in the proposed model.

8.4 SELF-WEIGHT RETAINING HEIGHT

In this section, the self-weight retaining height of unsaturated soil is considered as the first example of stability analysis with respect to unsaturated soil.

Considering the self-weight retaining height, the normal and tangential components of inter-particle stress vectors due to external force and seepage force in Equations (5.7.1) and (5.7.2) can be neglected. The following equations are then obtained:

$$F_{\beta,N} = F_{\beta,grav,N} + F_{\beta,matr,N} \tag{8.4.1}$$

$$F_{\beta,T} = F_{\beta,grav,T} \tag{8.4.2}$$

Substituting Equations (5.3.28), (5.3.29) and (5.5.20) into Equations (8.4.1) and (8.4.2), the following equations are obtained:

$$F_{\beta,N} = \frac{1}{1+e} \cdot (\rho_s - \rho_w) \cdot g \cdot D_{cha} \cdot \cos^2 \beta \cdot f_\beta(\beta) \cdot \beta$$
$$+ N_{ca,\beta} \cdot \left(2\pi \cdot T_s \cdot E[r_i'] + \pi \cdot s_u \cdot E\left[r_i'^2\right] \right) \tag{8.4.3}$$

$$F_{\beta,T} = \frac{1}{1+e} \cdot (\rho_s - \rho_w) \cdot g \cdot D_{cha} \cdot \sin\beta \cdot \cos\beta \cdot f_\beta(\beta) \cdot \beta \tag{8.4.4}$$

Considering the soil column with the height h_{self} which is divided into thin elements as shown in Figure 8.8, Equations (8.4.3) and (8.4.4) are rewritten as follows:

$$F_{\beta,N} = \frac{1}{1+e} \cdot (\rho_s - \rho_w) \cdot g \cdot D_{cha} \cdot \frac{h_{self}}{D_{cha}} \cdot \cos^2 \beta \cdot f_\beta(\beta) \Delta\beta$$
$$+ N_{ca,\beta} \cdot \left(2\pi \cdot T_s \cdot E[r_i'] + \pi \cdot s_u \cdot E\left[r_i'^2\right] \right)$$
$$= \frac{1}{1+e} \cdot (\rho_s - \rho_w) \cdot g \cdot h_{self} \cdot \cos^2 \beta \cdot f_\beta(\beta) \Delta\beta$$
$$+ N_{ca,\beta} \cdot \left(2\pi \cdot T_s \cdot E[r_i'] + \pi \cdot s_u \cdot E\left[r_i'^2\right] \right) \tag{8.4.5}$$

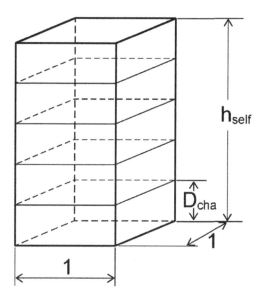

Figure 8.8 Soil column with unit cross-section area and h_{self}.

$$F_{\beta,T} = \frac{1}{1+e} \cdot (\rho_s - \rho_w) \cdot g \cdot D_{cha} \cdot \frac{h_{self}}{D_{cha}} \cdot \sin\beta \cdot \cos\beta \cdot f_\beta(\beta) \Delta\beta$$

(8.4.6)

$$= \frac{1}{1+e} \cdot (\rho_s - \rho_w) \cdot g \cdot h_{self} \cdot \sin\beta \cdot \cos\beta \cdot f_\beta(\beta) \Delta\beta$$

The inter-particle stress ratio $F_{\beta,T}/F_{\beta,N}$ is obtained as follows by using Equations (8.4.5) and (8.4.6):

$$\left(\frac{F_{\beta,T}}{F_{\beta,N}}\right)_{self} = \frac{\frac{1}{1+e} \cdot (\rho_s - \rho_w) \cdot g \cdot h_{self} \cdot \sin\beta \cdot \cos\beta \cdot f_\beta(\beta) \Delta\beta}{\frac{1}{1+e} \cdot (\rho_s - \rho_w) \cdot g \cdot h_{self} \cdot \cos^2\beta \cdot f_\beta(\beta) \Delta\beta + N_{ca,\beta} \cdot \left(2\pi \cdot T_s \cdot E[r_i'] + \pi \cdot s_u \cdot E[r_i'^2]\right)}$$

(8.4.7)

As h_{self} in Equation (8.4.7) is increased, $F_{\beta,T}/F_{\beta,N}$ also increases and is finally equal to $\tan\varnothing_{ave}$, i.e., the height h_{self} which satisfies the following equation and becomes the critical self-weight height h_{cri}.

$$\max\left(\frac{F_{\beta,T}}{F_{\beta,N}}\right)_{self} = \tan\varnothing_{ave} = \mu_{ave}$$

(8.4.8)

To obtain the critical self-retaining height h_{cri} mathematically becomes the extremal problem which can numerically be solved. The numerical procedure to obtain h_{cri} is shown in Figure 8.9.

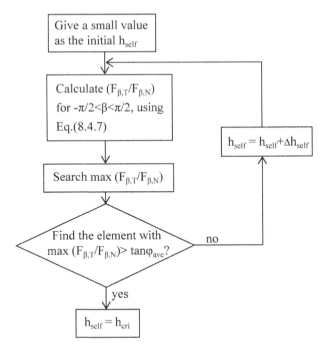

Figure 8.9 Calculation procedure of h_{cri}.

8.5 TYPICAL STABILITY ANALYSES IN GEOTECHNICAL ENGINEERING PROBLEMS BY PROPOSED MODEL

The Boussinesq's solution is applied to estimate the inter-particle stress vector due to external force in the ground, which means that the soil block is assumed to be an elastic body to estimate the inter-particle stress vector for the stability analysis.

8.5.1 Bearing capacity

Boussinesq (1885) derived the stresses for the semi-finite elastic body without the self-weight when the point load is applied. Figure 8.10 shows the point load P_{point} applied on the origin of the 3-dimensional orthogonal coordinates and the stresses generated the point (x,y,z), where the z-axis is taken in the vertical direction. The stresses derived by Boussinesq are shown for the 2-dimensional space as follows.

$$\sigma_x = \frac{3P_{point}}{2\pi} \cdot \left[\frac{z}{r^5} \cdot x^2 + \frac{1-2\nu}{3} \cdot \left\{ \frac{r^2 - rz - z^2}{r^3 (r+z)} - \frac{2r+z}{r^3 \cdot (r+z)^2} \cdot x^2 \right\} \right] \quad (8.5.1)$$

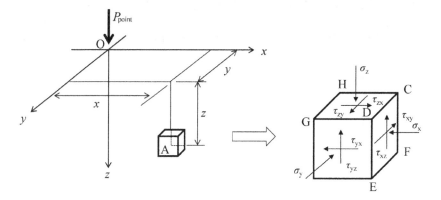

Figure 8.10 Stresses in the small element generated by point load.

$$\sigma_z = \frac{3P_{point}}{2\pi} \cdot \frac{z^3}{r^5} \tag{8.5.2}$$

$$\tau_{zx} = \frac{3P_{point}}{2\pi} \cdot \frac{z^2}{r^5} \cdot x \tag{8.5.3}$$

where

$$r^2 = x^2 + z^2 \tag{8.5.4}$$

ν:　　Poisson's ratio, which is not constant, but depends on strain and
P_{point}:　point load shown in Figure 8.10.

As the stresses shown in Equations (8.5.1)–(8.5.3) correspond to the inter-particle stress vectors due to the external force, the following equations are obtained, using Equations (5.6.11) and (5.6.12).

$$F_{\beta=\pm\frac{\pi}{2},ext,N} = \sigma_x \tag{8.5.5}$$

$$F_{\beta=0,ext,N} = \sigma_z \tag{8.5.6}$$

$$F_{\beta=\pm\frac{\pi}{2},ext,T} = \tau_{xz} \tag{8.5.7}$$

$$F_{\beta=0,ext,T} = \tau_{zx} \tag{8.5.8}$$

where

$$\tau_{xz} = \tau_{zx} \tag{8.5.9}$$

When the angle between the maximum principal stress plane and the horizontal plane is denoted by α, the relation between stresses of σ_x, σ_z, τ_{xz}

and τ_{zx} in Equations (8.5.5)–(8.5.8) on the Mohr's stress circle are shown in Figure 8.11. The normal and tangential components of inter-particle stress vector on the plane with the inclination angle β are then obtained as follows:

$$F_{\beta,\,\text{ext},N} = \frac{\sigma_z + \sigma_x}{2} + \cos\{2(\alpha + \beta)\}\cdot\sqrt{\left(\frac{\sigma_z - \sigma_x}{2}\right)^2 + \tau_{zx}{}^2} \tag{8.5.10}$$

$$F_{\beta,\text{ext},T} = \sin\{2(\alpha + \beta)\}\cdot\sqrt{\left(\frac{\sigma_z - \sigma_x}{2}\right)^2 + \tau_{zx}{}^2} \tag{8.5.11}$$

where

$$\tan 2\alpha = \frac{2\tau_{zx}}{\sigma_z - \sigma_x} \tag{8.5.12}$$

$F_{\beta,N}$ and $F_{\beta,T}$ in Equations (5.7.1) and (5.7.2) are rewritten to estimate the bearing capacity as follows, because the normal and tangential components of inter-particle stress vector due to seepage force is much too small:

$$F_{\beta,N} = F_{\beta,\text{grav},N} + F_{\beta,\text{mat},N} + F_{\beta,\text{ext},N} \tag{8.5.13}$$

$$F_{\beta,T} = F_{\beta,\text{grav},T} + F_{\beta,\text{ext},T} \tag{8.5.14}$$

Figure 8.12 shows the calculated domain of ground on which the small point load P_{point} acts. For simplicity, it is assumed that the ground consists

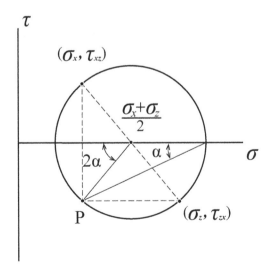

Figure 8.11 Stresses on Mohr's circle (Pole method).

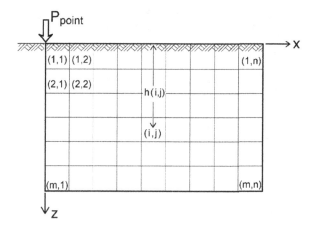

Figure 8.12 Small element (*m,n*) in the calculation domain for bearing capacity.

of coarse-grained soil with unique grain size distribution, void ratio and water content. The ground is divided by the small mesh which is named by the mesh number (i, j) $(i = 1 \sim m, j = 1 - n)$. The normal and tangential components of inter-particle stress vectors of the (i, j) mesh in Equations (8.5.13) and (8.5.14) are obtained as follows, except for the normal and tangential components of inter-particle stress vector due to external force, referring to Equations (5.3.28), (5.3.29) and (5.5.20):

$$F_{\beta,\mathrm{grav},N}\left(i,j\right) = \frac{1}{1+e}\cdot\left(\rho_s - \rho_w\right)\cdot g \cdot h_{(i,j)} \cdot \cos^2\beta\cdot f_\beta\left(\beta\right)\Delta\beta \qquad (8.5.15)$$

$$F_{\beta,\mathrm{grav},T}\left(i,j\right) = \frac{1}{1+e}\cdot\left(\rho_s - \rho_w\right)\cdot g \cdot h_{(i,j)} \cdot \sin\beta\cdot\cos\beta\cdot f_\beta\left(\beta\right)\Delta\beta \qquad (8.5.16)$$

where $h_{(i,j)}$: height of the (i,j) mesh shown in Figure 8.11.

$$F_{\beta,\mathrm{mat},N}\left(i,j\right) = N_{ca,\beta}\cdot\left(2\pi\cdot T\cdot E\left[r_i'\right] + \pi\cdot s_u\cdot E\left[r_i'^2\right]\right) \qquad (8.5.17)$$

Substituting Equations (8.5.10), (8.5.11), (8.5.15), (8.5.16) and (8.5.17) into Equations (8.5.13) and (8.5.14), $(F_{\beta,T}/F_{\beta,N})_{\mathrm{bear}}$ can be calculated.

$(F_{\beta,T}/F_{\beta,N})_{\mathrm{bear}}$ in the mesh number (i,j) are sequentially calculated, using Equations (8.5.13) and (8.5.14). As the point load P_{point} is increased, the mesh which satisfies the following equation begins to appear:

$$\max\left(\frac{F_{\beta,T}}{F_{\beta,N}}\right)_{\mathrm{bear}} \geq \tan\varnothing_{\mathrm{ave}} \qquad (8.5.18)$$

The point load at which the equation (8.5.18) is firstly satisfied is regarded as the bearing capacity.

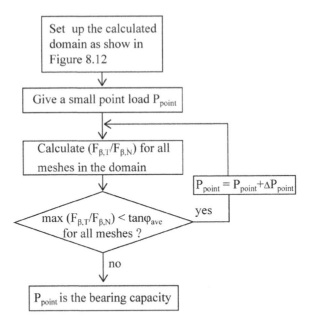

Figure 8.13 Calculation procedure to obtain bearing capacity.

Figure 8.13 shows the numerical procedure to obtain the bearing capacity.

8.5.2 Earth pressure

We can image the slope with the inclination angle $\theta = 90°$ and the height h as shown in Figure 8.14(a). Under this condition, the slope may be unstable when the height is larger than the self-weight retaining height h_{cri}. The pressure with the triangle distribution p_{earth} is applied to keep the slope stable as shown in Figure 8.14(b), i.e., p_{earth} regarded as the inter-particle stress vector due to external force is expressed as follows:

$$p_{earth} = \gamma_{earth} \cdot h \qquad\qquad (8.5.19)$$

where γ_{earth}: gradient of earth pressure with height as shown in Figure 8.14, corresponding to the unit weight of the soil block.

Rotating 90° of the point load, and then applying the Boussinesq's solution, as well as the bearing capacity and numerically integrating p_{earth}, the inter-particle stress vector due to external force in the mesh number (i, j) $(i = 1 \sim m, j = 1 - n)$ shown in Figure 8.14(b) is obtained. Then the same procedure as the bearing capacity is used to estimate the earth pressure, i.e., the normal and tangential components of inter-particle stress vector shown by Equations (8.5.13) and (8.5.14) are used to estimate stress condition of each mesh.

Figure 8.15 shows the calculation procedure to estimate the active and passive earth pressures. A small γ_{earth} in Equation (8.5.19) is given as the

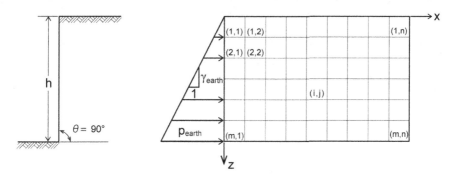

Figure 8.14 Small element (m,n) in the calculation domain for earth pressure.

earth pressure, and then the ratio of tangential component of inter-particle stress vector to normal component $(F_{\beta,T}/F_{\beta,N})_{earth}$ in the mesh number (i, j) $(i = 1 \sim m, j = 1 \sim n)$ is calculated according to similar procedures to obtain the self-weight height and bearing capacity. Initially, almost all meshes might satisfy $\max(F_{\beta,T}/F_{\beta,N})_{earth} \geq \tan\emptyset_{ave}$. As γ_{earth} is gradually increased, the meshes which satisfy $\max(F_{\beta,T}/F_{\beta,N})_{earth} < \tan\emptyset_{ave}$ are increased. If the lines which the meshes satisfied with $\max(F_{\beta,T}/F_{\beta,N})_{earth} < \tan\emptyset_{ave}$ $\max(F_{\beta,T}/F_{\beta,N})_{earth} < \tan\emptyset_{ave}$ can be connected in this process, the line are considered to be the slip line in the conventional soil mechanics. The stress condition under which all the mesh initially satisfies the following equation appears. This stress condition can be regarded as the active earth pressure condition.

$$\max\left(\frac{F_{\beta,T}}{F_{\beta,N}}\right)_{earth} < \tan\emptyset_{ave} \tag{8.5.20}$$

The active earth pressure $p_{earth,act}$ and its resultant force $p_{earth,act}$ are then expressed as follows.

$$p_{earth,act} = \gamma_{earth,act} \cdot h \tag{8.5.21}$$

where $\gamma_{earth,act}$: gradient of earth pressure at the active earth pressure condition.

$$P_{earth,act} = \frac{1}{2} \cdot p_{earth,act} \cdot h = \frac{1}{2} \cdot \gamma_{earth,act} \cdot h^2 \tag{8.5.22}$$

The pressure due to pore water should be added to the pressure due to soil particles except under dry conditions. Referring to Equation (5.3.36), the total earth pressure acted on the retaining wall is expressed as follows:

$$p_{total} = p_{earth,act} + p_{side} = \left(\gamma_{earth,act} + \frac{e \cdot S_r}{1+e} \cdot \rho_w g\right) \cdot h \tag{8.5.23}$$

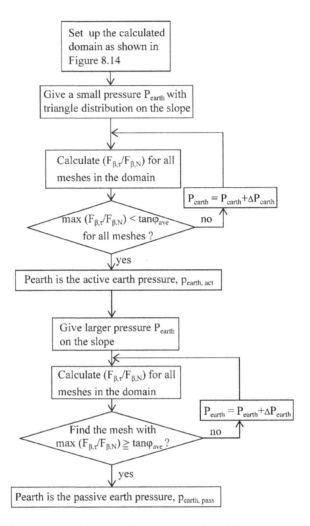

Figure 8.15 Calculation procedure to obtain active and positive earth pressure.

The resultant force is expressed as follows:

$$P_{total} = \frac{1}{2} \cdot \left(\gamma_{earth,act} + \frac{e \cdot S_r}{1+e} \cdot \rho_w g \right) \cdot h^2 \tag{8.5.24}$$

As γ_{earth} is furthermore increased, the meshes which satisfy the following equation begin to appear. If the line connected with these meshes can be drawn in this process, the line might be the slip line in the conventional soil mechanics.

$$\max \left(\frac{F_{\beta,T}}{F_{\beta,N}} \right)_{earth} \geq \tan\varnothing_{ave} \tag{8.5.25}$$

When Equation (8.5.25) is initially satisfies, p_{earth} is the passive earth pressure. Then the passive earth pressure $p_{earth,pass}$ and its resultant force $p_{earth,pass}$ are expressed as follows:

$$p_{earth,pass} = \gamma_{earth,pass} \cdot h \qquad (8.5.26)$$

where $\gamma_{earth,pass}$: gradient of earth pressure at the passive earth pressure condition.

$$P_{earth,pass} = \frac{1}{2} \cdot p_{earth,pass} \cdot h = \frac{1}{2} \cdot \gamma_{earth,pass} \cdot h^2 \qquad (8.5.27)$$

Similar to the active stress condition, the total earth pressure acting on the retaining wall under the passive stress condition is expressed as follows:

$$p_{total} = p_{earth,pass} + p_{side} = \left(\gamma_{earth,pass} + \frac{e \cdot S_r}{1+e} \cdot \rho_w g \right) \cdot h \qquad (8.5.28)$$

The resultant force is expressed as follows:

$$P_{total} = \frac{1}{2} \cdot \left(\gamma_{earth,pass} + \frac{e \cdot S_r}{1+e} \cdot \rho_w g \right) \cdot h^2 \qquad (8.5.29)$$

8.5.3 Slope stability

Analyzing the slope stability, the normal and tangential components of the inter-particle stress vector due to the external force and the seepage force in Equations (5.7.1) and (5.7.2) can be neglected.

Using Equations (5.3.28), (5.3.29) and (5.5.20), the following equations are obtained:

$$F_{\beta,N}(i,j) = \frac{1}{1+e} \cdot (\rho_s - \rho_w) \cdot g \cdot h_{(i,j)} \cdot \cos^2\beta \cdot f_\beta(\beta) \Delta\beta$$
$$+ N_{ca,\beta} \cdot \left(2\pi \cdot T \cdot E[r_i'] + \pi \cdot s_u \cdot E[r_i'^2] \right) \qquad (8.5.30)$$

$$F_{\beta,T}(i,j) = \frac{1}{1+e} \cdot (\rho_s - \rho_w) \cdot g \cdot h_{(i,j)} \cdot \sin\beta \cdot \cos\beta \cdot f_\beta(\beta) \Delta\beta \qquad (8.5.31)$$

where $h(i,j)$: height of the (i,j) mesh shown in Figure 8.16.

Figure 8.16 shows the slope with the inclination angle $\theta_{slope} < 90°$, the height h and the base length L_{base}, i.e., $\tan\theta_{slope} = h/L_{base}$. We can extend the procedure which derives the active earth pressure, applying the idea with respect to the counterweight fill method, i.e., replacing the distribution of active earth pressure by the soil block and the critical slope angle can be obtained in the following manner.

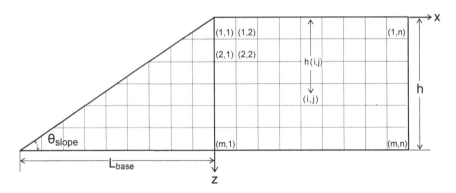

Figure 8.16 Small element (m,n) in the calculation domain for slope stability.

The weight of counterweight fill W is obtained as follows:

$$W = \frac{1}{2} \cdot \rho_t g \cdot L_{\text{base}} \cdot h \tag{8.5.32}$$

where ρ_t: wet density defined by Equation (3.1.11).

Applying the friction law for a particulate soil block expressed by Equation (8.1.5), the following equation is obtained for the critical slip condition of counterweight fill:

$$\mu_{\text{ave}} = \frac{\frac{1}{2}\gamma_{\text{earth}} \cdot h}{W} \tag{8.5.33}$$

Substituting Equation (8.5.32) into Equation (8.5.33), the following equation is obtained:

$$\mu_{\text{ave}} = \frac{\gamma_{\text{earth}}}{\rho_t g \cdot L_{\text{base}}} \tag{8.5.34}$$

Equation (8.5.34) is rewritten as follows.

$$L_{\text{base}} = \frac{\gamma_{\text{earth}}}{\mu_{\text{ave}} \cdot \rho_t g} \tag{8.5.35}$$

The ratio $\max(F_{\beta,T}/F_{\beta,N})_{\text{slope}}$ is calculated for the mech $(i,\ j)$ $(i=1\sim m,$ $j=1\sim n)$ in Figure 8.16 according to the similar procedure to calculate the active earth pressure. If all meshes do not satisfy the condition of $\max(F_{\beta,T}/F_{\beta,N})_{\text{slope}} < \tan\varnothing_{\text{ave}}$, the $\gamma_{\text{earth,act}}$ is increased until $\max(F_{\beta,T}/F_{\beta,N})_{\text{slope}} < \tan\varnothing_{\text{ave}}$ is satisfied for all meshes. The stress condition which firstly satisfies $\max(F_{\beta,T}/F_{\beta,N})_{\text{slope}} < \tan\varnothing_{\text{ave}}$ for all meshes is regarded as the critical state, i.e., the safety factor is 1. In this critical condition, the base length L_{cri} is calculated by Equation (8.5.35) and then the slope angle θ_{cri} is obtained as follows.

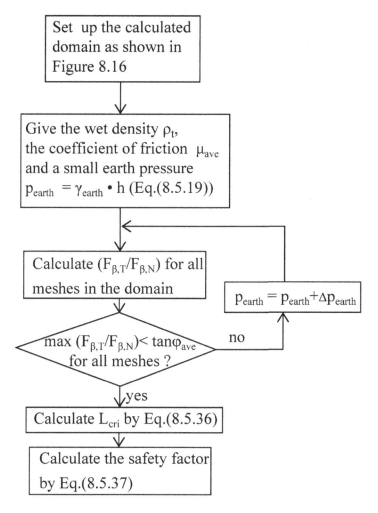

Figure 8.17 Calculation procedure to judge slope stability.

$$\theta_{cri} = \tan^{-1} \frac{h}{L_{cri}} \tag{8.5.36}$$

The safety factor is defined as follows:

$$F_s = \frac{L_{base}}{L_{cri}} \tag{8.5.37}$$

Figure 8.17 shows the calculation procedure to analyze the slope stability. The wet density ρ_t, the coefficient of average friction in a soil block μ_{ave} and the small tentative active earth pressure $p_{earth,act}$ are initially given. Then $(F_{\beta,T}/F_{\beta,N})$ is calculated by using Equations (8.4.1) and (8.8.4.2). The safety factor F_s of slope is calculated by giving the base length of slope L_{base} and using Equation (8.5.37).

As the normal component of an inter-particle stress vector due to the surface tension $F_{\beta,matr,N}$ is decreased by the change in water content with a change in weather such as rainfall, the number of meshes satisfied with $\max(F_{\beta,T}/F_{\beta,N}) \geq \tan\varnothing_{ave}$ is increased and the slip line might be drawn in some cases, using Equations (8.5.30) and (8.5.31) to calculate $F_{\beta,N}$ and $F_{\beta,T}$.

REFERENCES

Boussinesq, J. (1885). See Kimura, T. (1969). Stress propagation, Soil mechanics supervised by JSCE, Gihodo, 237–239 (in Japanese).

Fredlund, D. G., Rahardjo, H., and Fredlund, M. D. (2012). *Unsaturated Soil Mechanics in Engineering Practice*. John Wiley & Sons, 526–529.

Chapter 9

Deformation analysis using proposed models

9.1 MICROSCOPIC MOTION OF SOIL PARTICLES RELATING TO MACROSCOPIC DEFORMATION

The microscopic motion of soil particles which induces the observed macroscopic deformation is considered to consist of the continuous and discontinuous motions as shown in Figures 9.1 and 9.2.

Figure 9.1a and b schematically show the microscopic particulate soil structure for the coarse-grained soil at the stress state s and the stress state $s + \Delta s$ respectively, in which seven soil particles and a pore are included. Focusing the 7-th particle, the contact angles at two contact points change with the change in stress state from s to $s + \Delta s$ and it is found that these changes in contact angle contribute to the observed macroscopic deformation of soil block. The stress state s is evaluated with the normal and tangential components of inter-particle stress shown in Equations (5.7.1) and (5.7.2) in the proposed models.

Figure 9.2a, b and c schematically show the change in the particulate soil structure with the change in the stress state from s to $s + \Delta s$ and $s + 2\Delta s$ respectively. Figure 9.2a shows the particulate soil structure corresponding to Figure 9.1a. When the stress state changes from s to $s + \Delta s$, the 7-th soil particle drops into the pore due to the breakage of force equilibrium for the 7-th particle as shown in Figure 9.2b. When the stress state $s + \Delta s$ further proceeds to $s + 2\Delta s$, a renewed particulate soil structure is appeared as shown in Figure 9.2c.

The observed macroscopic deformation of soil specimen is considered microscopically to consist of the continuous and discontinuous motions of soil particles as shown in Figure 9.3, where the continuous and discontinuous motions respectively mean the change in contact angle as shown in Figure 9.1 and the disappearance and appearance of contact points as shown in Figure 9.2 during deformation process.

The continuous motion is evaluated by the change in contact angle at contact points of soil particles, i.e. the change in the probability density function for the contact angle, Equation (3.2.7), which will be discussed

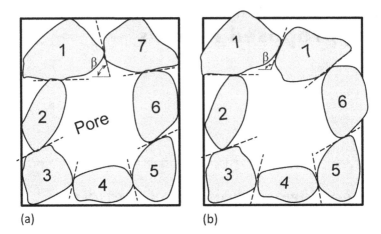

Figure 9.1 Continuous motion (change in contact angle) when the stress state changes from s to s + Δs (a) Stress state s, (b) Stress state s+Δs.

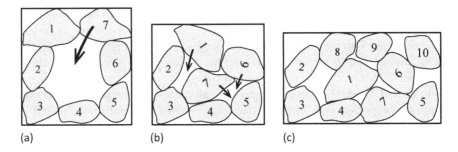

Figure 9.2 Discontinuous motions (disappearance and appearance of contact points) when the stress state changes from s to s + Δs and s + 2Δs (a) Stress state s, (b) Stress state s+Δs, (c) Stress state s+2Δs.

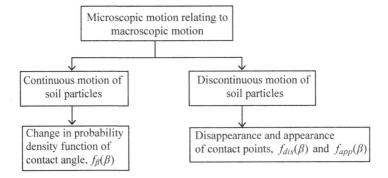

Figure 9.3 Microscopic components of deformation.

in Section 9.3.1. The discontinuous motion is evaluated by the ratio of the number of soil particles, which will be discussed in Section 9.3.2.

9.2 DERIVATION OF STRAIN INCREMENTS

Figure 9.4a shows a cuboid of soil block taken out of the triaxial specimen under the stress state s, corresponding to the cubic sample population in Figure 2.3. Figure 9.4b shows one of the paths connecting the soil particles between the bottom and top planes through the soil block, where the path is passed at the centroid of each soil particle. The height, width and depth of the cuboid are denoted by $L_{x1,s}$, $L_{x2,s}$ and $L_{x3,s}$ which are the lengths in the directions of the X_1-, X_2- and X_3-axes, where the direction of the X_1 axis is vertical and assumed to coincide with the principal strain increment in the triaxial test specimen. Denoting the distance between the j-the and the $(j+1)$-th particles in the direction of X_i-axis at the stress state s by $l_{x1,s,j}$, the following equation is obtained:

$$L_{x_i,s} = \sum_{j=1}^{N_{\text{path},X_i,s}} l_{x_i,s,j} \quad (i = 1,2,3) \tag{9.2.1}$$

where $N_{\text{path},X_i,s}$: number of soil particles in the path along the X_i-axis at the stress state s.

The principal strain (linear strain) increment in the direction of X_i-axis due to change in the stress state from s to $s + \Delta s$ is obtained as follows:

$$d\varepsilon_{X_i,s+\Delta s} = \frac{L_{X_i,s+\Delta s} - L_{X_i,s}}{L_{X_i,s}} \tag{9.2.2}$$

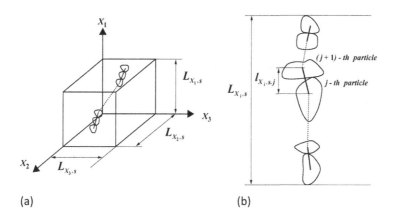

(a) (b)

Figure 9.4 Soil block with the lengths of $L_{xi,s}$ ($i = 1, 2, 3$) under stress state s (a) Soil block, (b) Path in soil block from bottom to top.

Substituting Equation (9.2.1) into Equation (9.2.2), the following equation is obtained:

$$d\varepsilon_{X_i,s+\Delta s} = \frac{\sum_{j=1}^{N_{path,X_i,s+\Delta s}} l_{X_i,s+\Delta s,j} - \sum_{j=1}^{N_{path,X_i,s}} l_{X_i,s,j}}{\sum_{j=1}^{N_{path,X_i,s}} l_{X_i,s,j}} \qquad (9.2.3)$$

Dividing both numerator and denominator by $N_{path,X_i,s}$ in Equation (9.2.3), the following equation is obtained:

$$d\varepsilon_{X_i,s+\Delta s} = \frac{\dfrac{1}{N_{path,X_i,s}}\sum_{j=1}^{N_{path,X_i,s+\Delta s}} l_{X_i,s+\Delta s,j} - \dfrac{1}{N_{prt,X_i,s}}\sum_{j=1}^{N_{path,X_i,s}} l_{X_i,s,j}}{\dfrac{1}{N_{path,X_i,s}}\sum_{j=1}^{N_{path,X_i,s}} l_{X_i,s,j}} \qquad (9.2.4)$$

In the following, the consideration is limited to the 2-dimensional space and then Equation (9.2.4) is rewritten as follows, using Equation (2.4.2):

$$d\varepsilon_{X_i,s+\Delta s} = \frac{\dfrac{N_{path,X_i,s+\Delta s}}{N_{path,X_i,s}} E\left[l_{X_i,s+\Delta s,j}\right] - E\left[l_{X_i,s,j}\right]}{E\left[l_{X_i,s,j}\right]} \qquad (i=1,3) \qquad (9.2.5)$$

where

$E\left[l_{X_i,s,j}\right]$:   mean length of adjacent soil particles projected on X_i-axis shown in Figure 9.4b and

$\dfrac{N_{path,X_i,s+\Delta s}}{N_{path,X_i,s}}$:   rate of change in number of soil particles included in the path of X_i-direction with change in the stress state from s to $s+\Delta s$.

Figure 9.5 shows the physical meaning of $N_{path,X_i,s+\Delta s}/N_{path,X_i,s}$ in Equation (9.2.5), where the number of particles included in the path shown in Figure 9.4b changes with the change in stress state from s to $s+\Delta s$. Therefore $N_{path,X_i,s+\Delta s}/N_{path,X_i,s}$ is the rate of change in number of soil particles included in the path with the change in stress state from s to $s+\Delta s$.

Using Equations (2.4.12), (3.2.6) and (3.2.7), $E\left[l_{X_1,s,j}\right]$ for X_1-axis $(i=1)$ and $E\left[l_{X_3,s,j}\right]$ for X_3-axis $(i=3)$ are obtained as follows:

$$E\left[l_{X_1,s,j}\right] = \int_{-\frac{\pi}{2}}^{\frac{\pi}{2}}\int_0^\infty D_s \cdot \cos\beta \cdot f_s(D_s) \cdot f_\beta(\beta;s)dD_s d\beta \qquad (9.2.6)$$

$$E\left[l_{X_3,s,j}\right] = \int_{-\frac{\pi}{2}}^{\frac{\pi}{2}}\int_0^\infty D_s \cdot \sin\beta \cdot f_s(D_s) \cdot f_\beta(\beta;s)dD_s d\beta \qquad (9.2.7)$$

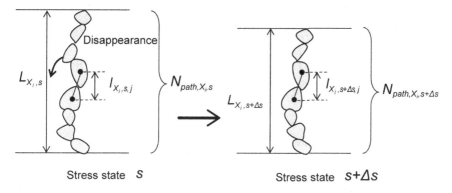

Figure 9.5 Discontinuous motion of soil particles (i.e., disappearance) with change in stress state from s to $s + \Delta s$.

where

$f_s(D_s)$: probability density function of diameter D_s (seeing Equation (3.2.6)),

$f_\beta(\beta; s)$: probability density function of contact angle β at the stress state s.

As the random variables D_s and β are independent of each other, Equations (9.2.6) and (9.2.7) are rewritten as follows, referring to Equation (2.4.7):

$$E\left[l_{X_1,s,j}\right] = \int_0^\infty D_s \cdot f_s(D_s)dD_s \cdot \int_{-\frac{\pi}{2}}^{\frac{\pi}{2}} \cos\beta \cdot f_\beta(\beta; s)d\beta = \bar{D}_s \cdot \int_{-\frac{\pi}{2}}^{\frac{\pi}{2}} \cos\beta \cdot f_\beta(\beta; s)d\beta \qquad (9.2.8)$$

$$E\left[l_{X_3,s,j}\right] = \int_0^\infty D_s \cdot f_s(D_s)dD_s \cdot \int_{-\frac{\pi}{2}}^{\frac{\pi}{2}} \sin\beta \cdot f_\beta(\beta; s)d\beta$$

$$= \bar{D}_s \cdot \int_{-\frac{\pi}{2}}^{\frac{\pi}{2}} \sin\beta \cdot f_\beta(\beta; s)d\beta \qquad (9.2.9)$$

where $\bar{D}_s$: mean diameter (which is different from D_{50}).

Substituting Equation (9.2.8) into Equation (9.2.5), the following equation is obtained in the direction of X_1-axis:

$$d\varepsilon_{X_1,s+\Delta s} = \frac{\dfrac{N_{path,X_1,s+\Delta s}}{N_{path,X_1,s}} \displaystyle\int_{-\frac{\pi}{2}}^{\frac{\pi}{2}} \cos\beta \cdot f_\beta(\beta; s + \Delta s)d\beta - \int_{-\frac{\pi}{2}}^{\frac{\pi}{2}} \cos\beta \cdot f_\beta(\beta; s)d\beta}{\displaystyle\int_{-\frac{\pi}{2}}^{\frac{\pi}{2}} \cos\beta \cdot f_\beta(\beta; s)d\beta} \qquad (9.2.10)$$

Substituting Equation (9.2.9) into Equation (9.2.5), the following equation is obtained in the direction of X_3-axis:

$$d\varepsilon_{X_1,s+\Delta s} = \frac{\dfrac{N_{path,X_1,s+\Delta s}}{N_{path,X_1,s}}\displaystyle\int_{-\frac{\pi}{2}}^{\frac{\pi}{2}}\sin\beta\cdot f_\beta\left(\beta;s+\Delta s\right)d\beta - \displaystyle\int_{-\frac{\pi}{2}}^{\frac{\pi}{2}}\sin\beta\cdot f_\beta\left(\beta;s\right)d\beta}{\displaystyle\int_{-\frac{\pi}{2}}^{\frac{\pi}{2}}\sin\beta\cdot f_\beta\left(\beta;s\right)d\beta} \tag{9.2.11}$$

9.3 EVALUATION OF CONTINUOUS AND DISCONTINUOUS MOTIONS OF SOIL PARTICLES

9.3.1 Continuous motion

The microscopic continuous motion of soil particles was explained in Section 9.1 where the continuous motion is defined as the change in contact angle with the change in stress state. It is assumed that the Markov chain explained in Section 2.7 can be applied to evaluate the change in contact angle quantitatively. The following equation is then obtained, using Equation (2.7.1).

$$f_\beta\left(\beta;s+\Delta s\right) = \sum_{\delta_\beta = -\frac{\pi}{2}-\beta}^{\frac{\pi}{2}-\beta}\left\{f_\beta\left(\beta-\delta_\beta;s\right)\cdot\Delta\beta\right\}\cdot P\left\{\beta\middle|\beta-\delta_\beta\right\} \tag{9.3.1}$$

where

$f_\beta(\beta;s+\Delta s)$: probability function of contact angle at stress state $s+\Delta s$,

δ_β: change in inclination angle β of contact plane with the change of stress state from s to $s+\Delta s$,

$f_\beta(\beta-\delta_\beta;s)\cdot\Delta\beta$: probability of inclination angle $\beta-\delta_\beta$ of contact plane,

$P\left\{\beta\middle|\beta-\delta_\beta\right\}$: conditional probability that the inclination angle $\beta-\delta_\beta$ at the stress state s changes to β at the stress state $s+\Delta s$.

Equation (9.3.1) is the basic equation of Markov chain for the random variable β in the deformation process.

Assuming that the probability density function of δ_β in Equation (9.3.1) can be expressed by the normal distribution with the mean value $\pm\Delta\beta_{ps}$ and the coefficient of variation κ_β, i.e., $N(\pm\Delta\beta_{ps},\kappa_\beta\cdot\Delta\beta_{ps})$, the following equation is obtained:

$$f_\delta\left(\delta_\beta\right) = \frac{1}{\kappa_\beta\cdot\Delta\beta_{ps}\sqrt{2\pi}}\exp\left\{-\frac{\left(\delta_\beta\pm\Delta\beta_{ps}\right)^2}{2\left(\kappa_\beta\cdot\Delta\beta_{ps}\right)^2}\right\} \tag{9.3.2}$$

where

$f_\delta(\delta_\beta)$:   probability density function of δ_β,

$\Delta\beta_{ps}$:   change in angle of potential slip plane (see Equations (8.2.1) and (8.2.2)) with the change of stress state from s to $s+\Delta s$,

κ_β: parameter to evaluate the variance of δ_β.

Equation (9.3.2) can be regarded as one of the constitutive Equation for the continuous motion of soil particles κ_β in Equation (9.3.2) is a fitting parameter at the present stage, but it is found from the numerical and experimental results shown in Section 10.5 that $\kappa_\beta = 1$ is the most optimum value and this value is then used in the calculation, i.e., the normal distribution of $N(\pm\Delta\beta_{ps}, \Delta\beta_{ps})$ is used for $f_\delta(\delta_\beta)$.

The potential slip plane was explained in Section 8.2 where the angle of the potential slip plane β_{ps} was defined as Equations (8.2.1) and (8.2.2) for the coarse-grained soil block. Figure 9.6 shows three states of contact angle β, i.e., in the first state, the contact angle β is larger than the angle of potential slip plane β_{ps} ($\beta > \beta_{ps}$); in the second state, the contact angle β is equal to the angle of potential slip plane β_{ps} ($\beta = \beta_{ps}$); and in the third state, the contact angle β is smaller than the angle of potential slip plane β_{ps} ($\beta < \beta_{ps}$).

Here we introduce the concept of a potential plane to estimate the change in contact angle with the stress state from s to $s+\Delta s$. The concept of a potential slip plane is that the probability of change in contact angel δ_β to be parallel to the potential slip plane is larger than the opposite change in contact angle when the stress state changes from s to $s+\Delta s$. Then three modes of change in the contact angle with the change in stress state are considered for the above three states respectively, i.e., the probability of a clockwise change in contact angle is larger than the anticlockwise change in the first

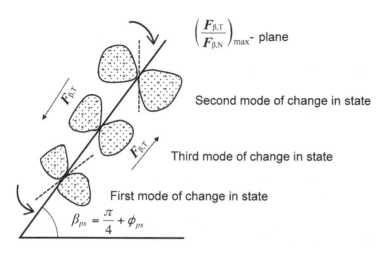

Figure 9.6 Concept of potential slip plane.

mode, the mean value of change in contact angle $E\left[\delta_\beta\right]$ at the contact angle $\beta = \beta_{ps}$ is equal to $\Delta\beta_{ps}$ in the second mode, and the probability of an anti-clockwise change in contact angle is larger than the clockwise change in the third mode. Based on the concept of a potential slip plane mentioned above, the following normal distribution with respect to δ_β can be derived:

For $0 \le \beta \le \beta_{ps}$, $-\dfrac{\pi}{2} \le \beta \le -\beta_{ps}$

$$f_\delta\left(\delta_\beta\right) = \frac{1}{\kappa_\beta \cdot \Delta\beta_{ps}\sqrt{2\pi}} \exp\left\{-\frac{\left(\delta_\beta - \Delta\beta_{ps}\right)^2}{2\left(\kappa_\beta \cdot \Delta\beta_{ps}\right)^2}\right\} \qquad (9.3.2(a))$$

For $-\beta_{ps} \le \beta \le 0$, $\beta_{ps} \le \beta \le \dfrac{\pi}{2}$

$$f_\delta\left(\delta_\beta\right) = \frac{1}{\kappa_\beta \cdot \Delta\beta_{ps}\sqrt{2\pi}} \exp\left\{-\frac{\left(\delta_\beta + \Delta\beta_{ps}\right)^2}{2\left(\kappa_\beta \cdot \Delta\beta_{ps}\right)^2}\right\} \qquad (9.3.2(b))$$

Using Equation (9.3.2), the conditional probability $P\{\beta|\beta - \delta_\beta\}$ in Equation (9.3.1) is expressed as follows:

$$P\{\beta|\beta - \delta_\beta\} = f_\delta\left(\delta_\beta\right) \cdot \Delta\delta_\beta \qquad (9.3.3)$$

Therefore Equation (9.3.1) is rewritten as follows:

$$f_\beta\left(\beta; s + \Delta s\right) = \sum_{\delta_\beta = -\frac{\pi}{2} - \beta}^{\frac{\pi}{2} - \beta} \left\{f_\beta\left(\beta - \delta_\beta; s\right) \cdot \Delta\beta\right\} \cdot f_\delta\left(\delta_\beta\right) \cdot \Delta\delta_\beta$$

$$= \sum_{\delta_\beta = -\frac{\pi}{2} - \beta}^{\frac{\pi}{2} - \beta} \left\{f_\beta\left(\beta - \delta_\beta; s\right) \cdot \Delta\beta\right\} \cdot \frac{1}{\kappa_\beta \cdot \Delta\beta_{ps} \cdot \sqrt{2\pi}} \cdot \exp\left\{-\frac{\left(\delta_\beta \pm \Delta\beta_{ps}\right)^2}{2\left(\kappa_\beta \cdot \Delta\beta_{ps}\right)^2}\right\} \cdot \Delta\delta_\beta \qquad (9.3.4)$$

Equation (9.3.4) is used to calculate the strains in Equations (9.2.10) and (9.2.11).

9.3.2 Discontinuous motion

The discontinuous motion of soil particle is defined and estimated as the disappearance and appearance of a contact point, as shown in Figure 9.5. Then $N_{path, X_i, s+\Delta s} / N_{path, X_i, s}$ in Equation (9.2.5) is one of the physical quantities to express the discontinuous motion due to the change in inter-particle stress condition. It is assumed that the discontinuous motion of disappearance and appearance of a contact point does not influence the continuous motion of change in the probability density function of contact angle. In other words, the probability density function of a contact angle's disappeared contact points is same as that of appeared contact points.

9.3.2.1 Disappearance

It is considered from Figure 9.5 that the probability of disappearance of the contact points is the maximum at the contact angle $\beta = \pm\pi/2$ (vertical contact plane) and the minimum at the contact angle $\beta = 0°$ (horizontal contact plane). The following normal distribution with the mean value $\pm\pi/2$ and the standard deviation $\kappa_{dis} \cdot \Delta\beta_{ps}$, i.e., $N(\pm\pi/2, \kappa_{dis} \cdot \Delta\beta_{ps})$ as shown Figure 9.7 can then be assumed to estimate the probability of disappearance:

$$f_{dis}(\beta) = \frac{1}{\kappa_{dis} \cdot \Delta\beta_{ps}\sqrt{2\pi}} \exp\left\{-\frac{\left(\beta \pm \frac{\pi}{2}\right)^2}{2\left(\kappa_{dis} \cdot \Delta\beta_{ps}\right)^2}\right\} \tag{9.3.5}$$

Equation (9.3.5) is rewritten as follows:

For $0 \le \beta \le \dfrac{\pi}{2}$

$$f_{dis}(\beta) = \frac{1}{\kappa_{dis} \cdot \Delta\beta_{ps}\sqrt{2\pi}} \exp\left\{-\frac{\left(\beta - \frac{\pi}{2}\right)^2}{2\left(\kappa_{dis} \cdot \Delta\beta_{ps}\right)^2}\right\} \tag{9.3.5(a)}$$

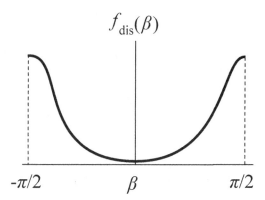

Figure 9.7 Probability density function for disappearance of contact points.

For $-\dfrac{\pi}{2} \le \beta \le 0$

$$f_{\text{dis}}(\beta) = \frac{1}{\kappa_{\text{dis}} \cdot \Delta\beta_{\text{ps}} \sqrt{2\pi}} \exp\left\{-\frac{\left(\beta + \dfrac{\pi}{2}\right)^2}{2\left(\kappa_{\text{dis}} \cdot \Delta\beta_{\text{ps}}\right)^2}\right\}$$ (9.3.5(b))

κ_{dis} in Equation (9.3.5) is a fitting parameter at the present stage, but it is found from the numerical and experimental results shown in Section 10.5 that $\kappa_{\text{dis}} = 1$ is the most optimum value and then the normal distribution $N\left(\pm\pi/2, \Delta\beta_{\text{ps}}\right)$ used for $f_{\text{dis}}(\beta)$ is used in the calculation.

Using Equation (9.3.5), the number of disappeared contact points per unit volume with the change in stress state from s to $s + \Delta s$ is obtained as follows:

$$N_{\text{cv,dis}} = \sum_{\beta=-\frac{\pi}{2}}^{\frac{\pi}{2}} \left(N_{\text{cv},\beta} \cdot f_{\text{dis}}(\beta)\Delta\beta\right)$$ (9.3.6)

where

$\quad N_{\text{cv,dis}}$: number of disappeared contact points per unit volume with the change in stress state from s to $s + \Delta s$ and

$\quad N_{\text{cv},\beta}$: number of contact points per unit volume with contact angle β which is obtained, using Equation (4.3.3) as follows:

$$N_{\text{cv},\beta} = N_{\text{cv}} \cdot f_\beta(\beta)\Delta\beta$$ (9.3.7)

9.3.2.2 Appearance

The dropping particle forms new contact points (i.e., appearance) under a new stress state. It is considered that the probability of appearance is maximum at the contact angle $\beta = 0°$ (horizontal contact plane) and minimum at the contact angle $\beta = \pm\pi/2$ (vertical contact plane). Then, assuming that the probability of disappearance is the normal distribution with the mean value 0 and the standard deviation $\kappa_{\text{app}} \cdot \Delta\beta_{\text{ps}}$ as shown in Figure 9.8, the following normal distribution, $N\left(0, \kappa_{\text{app}} \cdot \Delta\beta_{\text{ps}}\right)$ is obtained:

$$f_{\text{app}}(\beta) = \frac{1}{\kappa_{\text{app}} \cdot \Delta\beta_{\text{ps}} \sqrt{2\pi}} \exp\left\{-\frac{\beta^2}{2\left(\kappa_{\text{app}} \cdot \Delta\beta_{\text{ps}}\right)^2}\right\}$$ (9.3.8)

Using Equation (9.3.7), the number of appeared contact points per unit volume with the change in stress state from s to $s + \Delta s$ is obtained as follows:

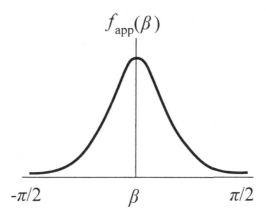

Figure 9.8 Probability density function for appearance of contact points.

$$N_{cv,app} = \sum_{\beta=-\frac{\pi}{2}}^{\frac{\pi}{2}} \left(N_{cv,\beta} \cdot f_{app}(\beta) \Delta\beta \right) \tag{9.3.9}$$

where $N_{cv,app}$: number of appeared contact points per unit volume with the change in stress state from s to $s + \Delta s$,

κ_{app} in Equation (9.3.7) is a fitting parameter at the present stage, but it is found from the numerical and experimental results shown in Section 10.5 that $\kappa_{app} = 1$ is the most optimum value and then the normal distribution $N(0, \Delta\beta_{ps})$ is used for $f_{app}(\beta)$ is used in the calculation.

9.3.2.3 Estimation of $N_{path,X_i,s+\Delta s}$ / $N_{path,X_i,s}$

$N_{path,X_i,s+\Delta s}/N_{path,X_i,s}$ in Equation (9.2.5) is defined as the rate of change in the number of soil particles included in the path of X_i-direction with the change in stress state from s to $s + \Delta s$. Here, assuming that the rate of change in the path of X_i –direction is same as in the path of all directions, i.e., the discontinuous motion in the soil block is isotropic, $N_{path,X_i,s+\Delta s}/N_{path,X_i,s}$ can be estimated by the change in the number of soil particles per unit volume which are estimated by the numbers of disappearance and appearance. Then the following equation is obtained, using Equations (9.3.6) and (9.3.9) (Kitamura et al., 2012).

$$\frac{N_{path,X_i,s+\Delta s}}{N_{path,X_i,s}} = \frac{N_{cv} - N_{cv,dis} + N_{cv,app}}{N_{cv}} \quad (i = 1, 3) \tag{9.3.10}$$

Substituting Equation (9.3.10) into Equations (9.2.10) and (9.2.11), the strain increment in the direction of X_i-axis $(i = 1, 3)$ with change in stress state from s to $s + \Delta s$ is obtained. It is found from Equation (9.3.10) that the discontinuous motion of disappearance and appearance does not depend on the direction of axis, i.e., the linear strain increment, although it is deeply related to the volumetric strain increment.

The settlement of coarse-grained soil due to submergence is called the collapsing and is considered to be one of the typical discontinuous behaviors. In the proposed models, the collapsing behavior can be expressed by the trigger of a sudden and extremely large decrease in the inter-particle stress $F_{\beta,\mathrm{matr},N}$ in Equation (5.7.1), i.e., the large change in β_{ps} defined by Equation (8.2.2) has suddenly occurred. And then $\Delta\beta_{\mathrm{ps}}$ in Equations (9.3.5) and (9.3.8) is increased, which brings out the increase in $N_{\mathrm{cv,dis}}$ in Equation (9.3.6) and $N_{\mathrm{cv,app}}$ in Equation (9.3.9).

9.3.3 Fitting parameters κ_β, κ_{dis} and $\kappa_{\beta\mathrm{app}}$

The values of fitting parameters κ_β in Equation (9.3.2), κ_{dis} in Equation (9.3.5) and κ_{app} in Equation (9.3.8) are assumed to be 1, which can be selected as the optimum value from the comparison of numerical simulation with triaxial soil test results in the present stage (Kitamura et al., 2012). But more soil tests on various soils and their numerical simulations should be carried out to check the validity of this value. The applying of data assimilation is considered to be one of the promising methods to estimate these fitting parameters, using the data obtained from the triaxial shear compression tests. The authors are now composing a computer program for the data assimilation, although the results are not shown in this chapter.

REFERENCE

Kitamura, R., Yamada, M., Kawabata, K., Inagaki, Y. and Araki, K. (2012). Mechanical and numerical model for deformation behavior of unsaturated soil. *Journal of Japan Society of Civil Engineers*, Division I, 68(2), I_487–I_492 (in Japanese).

Chapter 10

Numerical simulation for saturated–unsaturated soil tests

The numerical simulation is carried out by the proposed model for Shirasu which is a popular volcanic soil and was sampled from Sendai River in Kagoshima Prefecture, Japan. The numerical results of the soil water characteristic curve, the coefficient of water permeability and the self-weight height are shown in this chapter.

10.1 FUNDAMENTAL PHYSICAL QUANTITIES OF SHIRASU

Table 6.2 shows the soil properties of Shirasu used in the numerical simulation, i.e., the density of soil particles and the void ratio. The density of Shirasu is smaller than common sandy soil because Shirasu particles consists of porous glass material. Table 6.1 shows the percentage of fineness by weight obtained from the sieve and sedimentation analyses for Shirasu.

Figures 6.12(b) and 7.6(b) show the corrected pore size distributions used for the soil water characteristic curve and the coefficient of water permeability respectively. The parameters of the original and corrected pore size distributions for the soil water characteristic curve and the coefficient of water permeability are listed in Table 10.1.

10.2 NUMERICAL SIMULATION

10.2.1 Soil water characteristic curve

Figure 10.1 shows the original and corrected soil water characteristic curves with the plots obtained from the water retention soil test. It is found that the soil water characteristic curve can be estimated by using the grain size distribution and void ratio, and a more accurate curve can be obtained when the data (w, s_u) from the water retention soil test are known, where w is the water content and s_u is the suction.

Table 10.1 Parameters of Original and Corrected Pore Size Distributions

Distribution parameters	Symbols	Original	Corrected for soil-water characteristic curve	Corrected for coefficient of water permeability
Mean value of $\ln D_v$	λ_v	−5.128	−6.338	−10.264
Standard deviation of $\ln D_v$	ζ_v	2.046	2.046	2.0462
Mean value of D_v [m]	μ_v	4.81×10^{-4}	1.43×10^{-4}	2.83×10^{-6}
Standard deviation of D_v [m]	σ_v	3.87×10^{-3}	1.15×10^{-3}	2.28×10^{-5}
Coefficient of variation	κ	8.051	8.051	8.051

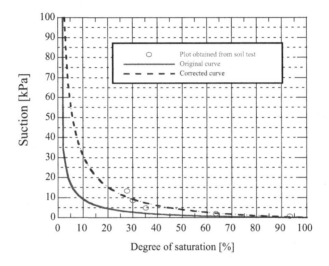

Figure 10.1 Original and corrected soil water characteristic curve with measured plots.

10.2.2 Coefficient of water permeability

Figure 10.2 shows the original and corrected water permeability functions with the plots obtained from the water permeability soil test. It is found that the water permeability function can be estimated by using the grain size distribution and void ratio and the more accurate curve can be obtained when the data (w, k_w) from the water permeability soil test are known, where w is the water content and k_w is the coefficient of water permeability.

10.2.3 Self-weight retaining height

According to the procedure shown in Figure 8.9, the self-weight retaining height can be calculated by using the following equation:

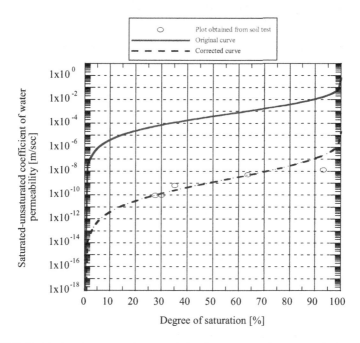

Figure 10.2 Original and corrected permeability functions with measured plots.

$$\left(\frac{F_{\beta,T}}{F_{\beta,N}}\right)_{self}$$

$$= \frac{\dfrac{1}{1+e} \cdot (\rho_s - \rho_w) \cdot g \cdot h_{self} \cdot \sin\beta \cdot \cos\beta \cdot f_\beta(\beta)\Delta\beta}{\dfrac{1}{1+e} \cdot (\rho_s - \rho_w) \cdot g \cdot h_{self} \cdot \cos^2\beta \cdot f_\beta(\beta)\Delta\beta + N_{ca,\beta} \cdot \left(2\pi \cdot T_s \cdot E[r_i'] + \pi \cdot s_u \cdot E[r_i'^2]\right)} \quad (8.4.7bis)$$

where

$$f_\beta(\beta) = \pm\frac{2/\pi - 2\cdot\varsigma_c}{\pi/2}\cdot\beta + \frac{2}{\pi} - \varsigma_c \quad (3.2.7bis)$$

$$N_{ca,\beta} = N_{ca}\cdot f_\beta(\beta)\Delta\beta \quad (5.2.9bis)$$

$$E[r_i'] = \int_0^\infty \frac{-3T_s + \sqrt{9T_s + 4D_s\cdot s_u\cdot T_s}}{2s_u}\cdot f_s(D_s)dD_s \quad (5.5.23bis)$$

$$E[r_i'^2] = \int_0^\infty \left(\frac{-3T_s + \sqrt{9T_s + 4D_s\cdot s_u\cdot T_s}}{2s_u}\right)^2\cdot f_s(D_s)dD_s \quad (5.5.24bis)$$

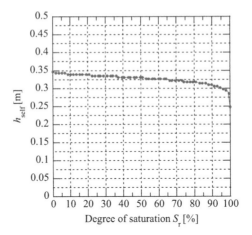

Figure 10.3 Relation between self-weight retaining height and degree of saturation.

Figure 10.3 shows the calculated relation between the self-weight retaining height and the degree of saturation, where the degree of saturation is calculated by using the corrected pore size distribution shown in Figure 7.6(a) and the corrected soil water characteristic curve shown in Figure 10.2. It is found that the self-weight retaining height is decreased with the increase in the degree of saturation.

Chapter 11

Issues to be solved in future

An original, intrinsic and versatile mechanical model for coarse-grained soil is proposed in this book, aiming for a paradigm shift with respect to the discipline of soil mechanics. The microscopic considerations are done to analyze the motion of soil particles and the predominant flow direction of pore water and pore air in the soil block. The motion of soil particles is estimated by the change in the contact angle at the contact point (continuous motion) and the disappearance and appearance of contact points (discontinuous motion). The flow of pore water and air is estimated by the elementary particulate model (EPM), then the mechanical behaviors of soil particles and pore water and air are modeled, aided by probability theory and statistics. The fundamental background concepts supported in this book are that the macroscopic physical quantities measured by conventional laboratory tests and in-situ tests are the mean value of the microscopic physical quantities.

The proposed model can be used to analyze the mechanical behaviors of coarse-grained soil in the research field of soil mechanics, to predict soil test results such as the soil water characteristic curve and the water permeability function which are time- and cost-consuming tests in the practical field of geotechnical engineering. On the other hand, the proposed model can be refined to verify the validity of the laboratory and in-situ soil tests and various observed data in the field, by means of the learning effect.

However, at the present stage, it is difficult to measure the microscopic physical quantities quantitatively using current technology. Several assumptions are included in the proposed model which are listed in Table 11.1.

The first assumption with respect to the grain size distribution function:

> The grain size accumulation curve obtained from the grain size analysis is modeled by the logarithmic normal distribution in the proposed model. The authors think that it might be valid for coarse-grained soil, but occasionally invalid for fine-grained soil. However, if the grain size accumulation curve can be expressed by the other distribution function, the proposed model is applicable, replacing the logarithmic normal distribution by the other distribution function.

Table 11.1 Several assumptions are included in the proposed model.

	Items	Assumptions
1	Grain size distribution	The grain size distribution is assumed to be the logarithmic normal distribution.
2	Pore size distribution	The pore size distribution is assumed to be the logarithmic normal distribution, as well as the grain size distribution. The coefficient of variance is also the same as that of grain size distribution.
3	Shape of soil particles	The shape of soil particles is assumed to be spherical.
4	Maximum void ratio	The maximum void ratio in EPM is assumed to be 3.66, although there might be samples (EPB) with larger void ratio in the sample population.
5	Number of contact points per soil particle	The empirical equation proposed by Fields (1963) is adopted to estimate the number of contact points per soil particle.
6	Distribution of contact angle	The shape of distribution of contact angle is assumed to be pentagonal and the ratio of minimum height to maximum height of pentagon is 1:3, based on the experimental results by Oda (1972).
7	Distribution of inclination angle in EPM	The distribution of the inclination angle of pipe in EPM which indicates the direction of predominant flow in EPB is assumed to be same as that of the contact angle.
8	$\mu_{ave} = \tan\phi_{repose}$	The coefficient of average friction in the soil block μ_{ave} is assumed to be same as $\tan\phi_{repose}$, where ϕ_{repose} is the angle of repose of coarse-grained soil.
9	Inter-particle stress due to external force	The Boussinesq's solution is applied to estimate the inter-particle stress due to external force. It means that the soil block is assumed to be an elastic body.

The second assumption with respect to the pore size distribution function:

It is assumed that the pore size distribution function is the logarithmic normal distribution and its coefficient of variation is same as that of the grain size distribution function. The validity of this assumption will be made clear when the pore size distribution can be directly measured by some devices. The logarithmic normal distribution will then be replaced by the measured pore size distribution function.

The third assumption with respect to the shape of soil particles:

It is assumed that the shape of soil particles is spherical as shown in Figure 3.9, which is related to how the diameter of soil particles obtained from the grain size analysis is estimated by assuming that the shape of soil particles is spherical. As it is anticipated that the mean shape of irregular soil particles in the soil block is spherical, the authors think that this assumption might be acceptable.

The fourth assumption with respect to the void ratio derived from EPM:

> The maximum void ratio derived from EPM is assumed to be 3.66, although there might be a few samples (EPBs) with larger void ratio in the sample population.

The fifth assumption with respect to the number of contact points per soil particle:

> The number of contact points per soil particle is estimated for coarse-grained soil by the empirical equation proposed by Field (1963) in the proposed model. When the method to estimate the number of contact points can be established by using results from computer technology such as CT and image-processing, the empirical equation will be replaced by more accurate one.

The sixth assumption with respect to the distribution function of contact angle:

> The distribution function of contact angle is assumed to be a pentagonal in shape in the proposed model. When the method to measure the contact angle can be established by using the results of computer technology, the pentagonal distribution function will be replaced by more accurate one.

The seventh assumption with respect to the inclination angle in EPM:

> The distribution function of inclination angle in EPM is assumed to be same as that of contact angle in the proposed model. When the method to measure the predominant flow direction can be established by using the results of computer technology, the distribution function of predominant flow direction will be replaced by more accurate one.

The eighth assumption with respect to the coefficient of average friction in the soil block:

> The coefficient of average friction in a soil block μ_{ave} is assumed to be same as $\mu_{ave} = \tan\varnothing_{repose}$, where $\varnothing_{repose}$ is the angle of repose and corresponds to the internal friction angle in the conventional soil mechanics. The validity of this assumption will be checked by many numerical simulations on various soils in the near future.

The ninth assumption with respect to the estimation of inter-particle stress due to an external force and/or pressure:

> Boussinesq's solution is applied to estimate the inter-particle stress due to an external force and/or pressure. It means that the soil block is

assumed to be an elastic body, although the soil block is a particulate material and not an elastic body. As the first stage, the inter-particle forces at contact points should be measured for the soil block by using the computer and electronic technology, such as CT and image-processing, that are developed in the near future (it might be effective to measure the inter-particle force in the cosmic space if possible). Then the measured inter-particle forces should then be related to the inter-particle stress. Finally the validity of Boussinesq's solution will be verified.

REFERENCES

Field, W. G. (1963). Towards the statistical definition of a granular mass. Proceedings of 4th A and N.Z. Conf. on SMFE, 143–148.
Oda, M. (1977). Co-ordination number and its relation to shear strength of granular material. *Soils and Foundations*, 17(2), 29–42.

Index

Printed in the United States
by Baker & Taylor Publisher Services